No Place Like Home

No Place Like

Home

A HISTORY OF

NURSING AND HOME CARE

IN THE

UNITED STATES

Karen Buhler-Wilkerson

THE JOHNS HOPKINS UNIVERSITY PRESS

BALTIMORE AND LONDON

© 2001 The Johns Hopkins University Press
All rights reserved. Published 2001
Printed in the United States of America on acid-free paper

Johns Hopkins Paperbacks edition, 2003
2   4   6   8   9   7   5   3   1

The Johns Hopkins University Press
2715 North Charles Street
Baltimore, Maryland 21218-4363
www.press.jhu.edu

Library of Congress Cataloging-in-Publication Data

Buhler-Wilkerson, Karen.
No place like home : a history of nursing and home care
in the United States / Karen Buhler-Wilkerson.
p. cm.
Includes bibliographical references and index.
ISBN 0-8018-6598-0 (hardcover : alk. paper)
1. Home nursing—History. I. Title.
RT120.H65 B84 2001
362.1′4′0973—dc21
00-010275

ISBN 0-8018-7318-5 (pbk.)

A catalog record for this book is available from the British Library.

Contents

Preface and Acknowledgments

Although we were neighbors and occasionally exchanged small talk on the street, until three years ago I barely knew Henry Howard (not his real name). It is difficult now to imagine how very little I knew about him. Likewise, he did not know me, nor did he know I was a nurse. When he rang the doorbell on a February day in 1997, he had just been discharged from the hospital following surgery for esophageal cancer. He could not speak and communicated only by written messages, and he clearly needed help with his feeding tube, tracheostomy, and other health care problems. For the next year I took on the role of good neighbor and caregiver and intimately experienced home care the American way.

Without question, Henry remained the central actor in this story until he died. He was fearless in his orchestration of events and people, and his tenacity was admirable, even if at times wearing in its demands and seemingly unrealistic expectations. Whether negotiating with home care workers, his medical oncologist or surgeons, staff in the emergency room, nurses and social workers, a difficult landlord, an attorney, or a funeral director, Henry remained quintessentially true to an inner dynamic characterized by independence, stubbornness, and originality.

Tough as it was, it was his life, and I was only trying to help. Assistance included looking for solutions to needs for communication, nutrition, elimination, medication, and housekeeping and referral to other experts as necessary. Many of our experiences revealed a fragmented system of care. Henry and I regularly dealt with the problems associated with absence of a single provider or coordinator of care and with maddening reimbursement structures, despite a respectable insurance plan. In the end, the only way to get the necessary care, scheduled when most needed and in an appropriate amount, was for Henry to pay for it out of his own limited and carefully guarded resources. Unfortunately, most of us could not afford such costs.

Over the course of Henry's illness, activities such as moving to a new apartment, selecting a more comfortable easy chair, shopping for and preparing food, and walking the dog were transformed from routine events to undertakings of crisis proportions. Helping with the simple acts of daily living, along with managing the various technologies now available for home care, providing transportation to numerous appointments, and filling prescriptions for drugs and eyeglasses, took more time than my academic schedule permitted—but somehow these tasks were done. Throughout it all, the single most essential participant was Cheryl McEntee, "our nurse" from the Visiting Nurse Association of Greater Philadelphia. She arrived confident, experienced, smart, and caring. As Henry's medical complications intensified, she was there. We trusted her and counted on her to manage both the care and us. After Henry died, she too grieved for this stranger who had shown up at the door. Despite her ingenuity and resourcefulness and my thirty years of experience in home care, coverage limitations and eligibility criteria made it impossible to provide the services that Henry required to remain at home. Almost twelve months after his saga began, Henry died alone in the hospital following an acute episode that occurred when no one was available to help.

A unique and intense year brought challenges and opportunities. From Henry's final journey I learned a great deal more about home care. While patterns of home care delivery remain confusing to both consumers and providers, the idea of sending expert caregivers into the homes of the sick and vulnerable retains its special appeal as practical, necessary, and family-oriented care. In my examination of the history, public policy, and personal consequences of caring for the sick at home, I have remained convinced that a caring society should choose to invest in a more coordinated and comprehensive approach. Instead, we remain ambivalent about whether caring for the sick at home is a family responsibility or one to be shared by the larger community. We understand the problems, know the answers, and have the resources, but we lack the public mandate to change how we care for the sick at home.

In this book, I offer an analysis of the history of home care from the 1880s to the enactment of Medicare legislation in 1965. Specifically, it is the history of organized home care. As such, it provides occasional glimpses of the equally important and complex subject of family caregiving. Like hospital care, home care replicates fundamental social relationships and values. As the antithesis of institutional care, however, its clear and consistently understood vision is focused on the home. The central question of this book is why, despite its potential as a preferred, rational, and possibly cost-effective alternative to institutional care, home care remains a marginalized experiment in caregiving.

Whether in the sickrooms of the rich, the middle class, or the poor, this book examines how race, ethnicity, income, gender, type of illness, local conditions, and patterns of practice have influenced access to and quality of care. From a new perspective, this study illuminates evolving views on family, home, disease, dependency, science, and technology. It places local ideas about the formation and function of home-based services within a framework that links them to national events and health care agendas. I give special attention to care of the "dangerous sick," particularly poor immigrants with infectious diseases, and the "uninteresting sick," people with chronic illnesses. I explore the interactions between philanthropic and entrepreneurial home care, as well as how the endless obligation of caring for the sick at home is viewed from the perspective of nurses, physicians, and family caregivers. Recalling home care's reformulation over the past one hundred years, this is also a study of ambiguous attitudes in the United States about individual versus community responsibility for the social and health needs of the sick. Finally, this book evaluates the impact of social attitudes, medical advances, demographic changes, and economic factors in shifting the locus of care, first from home to hospital and, more recently, back to the home again. From this layered story we begin to understand the situational complexities encountered by doctors, nurses, and families caring for the sick at home.

The book is divided into four chronological parts. The first details the origins of organized home care, from its beginnings in Charleston, South Carolina, to its development on a much larger scale in urban areas on the cusp of the twentieth century. I also consider the role of philanthropy and the evolving roles of patrons and professionals. These chapters chronicle the transfer of leadership from "ladies," who managed all the work, to nurses, and the resulting professionalization of care of the sick at home. In the next part I examine the work and reality of caring for the sick at home, including care of the acutely, chronically, and "dangerously" ill. The effects of ethnicity, race relations, and social class on home care are also examined. The last chapter in Part II reviews nurse-visionary Lillian Wald's invention of public health nursing as a new paradigm for community-based nursing practice within the context of social reform.

Part III considers the relationship between mission and finances in the business of nursing at home. I explore the impact of insurance coverage on care of the sick at home between 1909 and 1952 and offer an analysis of the economic ramifications of delivering home care in a rapidly changing world. This part of the book also considers care of the sick as a business over one specific forty-year period when most nurses found employment in private homes, as well as the shift of private-duty nurses from home care to hospital work. Part IV takes a more recent view of home care's mission and purpose and the familiar themes of chronic illness, hospital dominance, financial viability, and struggles to survive. This part closes with the inclusion of home care benefits in the Medicare program in 1965. In the Epilogue I explore the relationship between historical events and the projected future of home care.

Although often invisible and unrecognized, the problem of caring for the sick at home is familiar to most Americans. It involves the perspectives of many actors, from beleaguered family caregiver to underpaid health care worker to besieged agency administrator. For more than a century, communities in the United States have relied on a complex assortment of private, voluntary, and public organizations to supplement family caregiving. Because women are so often either the patients or the caregivers and managers of home care, this history reflects the lives of many women. More than seven million Americans currently receive home care, but recent government payment reforms threaten to jeopardize eligibility for and access to needed services. Today, twenty thousand home care agencies and their staffs of nurses and other home care workers are confronted in very pressing ways by difficult questions of how much care society is willing to pay for, who should receive that care, who should provide it, and for how long. Likewise, family caregivers struggle to find ways to cope with their daily dilemmas, challenges, and circumstances. As Ruth Hubbard, General Director of the Philadelphia Visiting Nurse Society, concluded fifty years ago, the history of home care provides the backdrop against which its future will develop. The permanency of the challenge and the inevitability of change provide home care's enduring foundation.[1]

The conclusion to what sometimes seemed, to me and those around me, an endless project is indeed a relief. Thankfully, most friends, colleagues, editors, and family members remain with me. I, and they, look forward to life following the "birth" of this book. Nevertheless, it brings great satisfaction to at last thank an extraordinary group of people, each of whom contributed in immeasurable ways to making the book a reality.

I begin with my father, John E. Buhler, an exceptional person who loved, nurtured, taught, and established great expectations. Dean of two dental schools from the age of forty onward, he was the consummate academic. An early historian of dentistry, he stimulated my interest in history. When I was only eleven years old, he sent me a note, one I have since treasured, wishing me a "confident, useful, and satisfying life." That message, written in 1955 during my southern childhood, has sustained me all of my life, and most especially during the writing of this book. Believing that there was no place like home, my father chose to die at home. His care was provided by my mother, Ruth Buhler, an expert nurse and skillful family caregiver. Thank you Dad—this one is for you.

Many debts accrued over a decade of research and writing. I am particularly appreciative of collaborative work with Audrey Davis, at that time with the Smithsonian Institution, on the history of public health nursing. We traveled many miles together, visiting archives in New York, Cleveland, and Chicago. I hope this book is a satisfying conclusion to that journey. I thank Sam Fagin for the book's title, spontaneously produced on a beautiful afternoon with his wife, Claire Fagin, sitting in their living room and gazing out on Central Park. My editor, the divine Mary Norris, has been patient and persistent, uncovering errors in footnotes, correcting punctuation, and always choosing the right word. The editorial staff at the Johns Hopkins University Press, especially Jacqueline Wehmueller, was most helpful. Betsy Weiss, administrative assistant at the Center for the Study of the History of Nursing at the University of Pennsylvania, has been supportive in so many ways. Elizabeth Logue provided invaluable research assistance, persistently reading "miles" of microfilm. I give special thanks to Lynn Brown for many years of insights and wise counsel. My sons, David and Jonathan, and Phil Lavene, the School of Nursing's computer wizard, mercifully maintained an aging and often cranky computer and managed my technophobia. Over the years, my graduate students have taught me a great deal about the history of nursing and health care. My dog, Polly, and five cats (La Nuit, Smokey, Mabel, Lavinia, and Isabel) walked over, sat on, and rearranged many versions of this manuscript. I thank you all.

Research for this book was made possible by the attentive efforts of the librarians and archivists at the Special Collections of the New York Public Library; Medical Archives of New York Hospital Cornell Medical Center; Historical Collections, College of Physicians of Philadelphia; Schlesinger

Library, Radcliffe College; Sterling Memorial Library Archives, Yale University; Nursing Education Archives, Teachers College, Columbia University; Rare Book and Manuscript Library, Columbia University; Metropolitan Life Insurance Company Library and Archives; Rockefeller Archives and Research Center; National Archives; Western Reserve Historical Society; Chicago Historical Society; South Carolina Historical Society; Waring Library of the Medical University of South Carolina; Hampton University Archives; Tompkins-McCaw Library, Richmond, Virginia; and the Center for the Study of the History of Nursing at the University of Pennsylvania School of Nursing. Numerous visiting nurse associations have kindly allowed me to examine their collections. Special thanks go to the Visiting Nurse Service of New York, Visiting Nurse Association of Greater Philadelphia, Cleveland Visiting Nurse Association, Chicago Visiting Nurse Association, Ladies Benevolent Society of Charleston, South Carolina, and the Providence, Rhode Island, District Nurses Association. I am also indebted to Mrs. Dorothy Anderson, whose numerous Charleston connections, introductions, and insights were invaluable.

This project has been generously supported by fellowships from the National Library of Medicine (grant no. RO1 LM06304-01); National Center for Nursing Research (grant no. NR-02078); University of Pennsylvania Research Foundation; University of Pennsylvania Boettner Financial Gerontology Research Fund; and Frank Morgan Jones Fund at the University of Pennsylvania School of Nursing. I give very special thanks to Mark Baiada and Bayada Nurses Home Care Specialists, with whom I have been closely associated for over fifteen years. Mark Baiada has enthusiastically sponsored this project and throughout it all has been a wise tutor and a great champion of nursing and home care.

I am also grateful to my friends and colleagues who have read chapters in progress, as well as the completed manuscript. Their thoughtful suggestions, numerous edits, and remarkable stamina—beyond the call of duty—made this book possible. My special thanks go to Susan Reverby, Joan Lynaugh, Charles Rosenberg, Vanessa Gamble, Gretchen Condran, Barbara Bates, and Barbara Savage. I am especially indebted to my partner in life, Neville E. Strumpf, who painstakingly read, questioned, or modified nearly every sentence (including this one) and made this a better book. Any remaining errors are mine.

Finally, to all who care for the sick at home—thank you. There *is* no place like home!

No Place Like Home

My view you know,

is that the ultimate destination of all nursing

is the nursing of the sick in their own homes . . .

I look to the abolition of all hospitals . . .

But no use to talk about the year 2000.

FLORENCE NIGHTINGALE

JUNE 1867

Prelude

*I*n early nineteenth-century America, care for the sick was part of domestic life, guided by family traditions and the advice found in the medical or nursing manuals of the day. When families hired physicians or nurses, professional care was delivered in the patient's home, most often with the assistance of female family members, neighbors, or occasionally a servant. Yet for those without family or financial resources, few health care options were available. This chapter examines how one charitable group, the Ladies Benevolent Society of Charleston, South Carolina, addressed these problems. The story of this society provides insight into the complex issues encountered by home care providers for nearly two hundred years: family circumstances, chronic disease, race, poverty, and "difference."

The Ladies of Charleston, South Carolina

Similar in many ways to other antebellum cities, Charleston nevertheless had a unique racial and class composition that posed significant challenges as members of the Ladies Benevolent Society (LBS) worked to fulfill the society's mission and social responsibilities. Rather than disappearing after the Civil War, these differences expanded and intensified to include the fear of "others" as potentially menacing or dangerous. In this "queen city of the south," in the early nineteenth century, a group of affluent women were the first to confront and document the dilemma of caring for the sick at home.[1]

Every city and community in the early 1800s was confronted with questions associated with care of sick people: Who were the worthy recipients of

1

care? What amount, type, and length of care was most appropriate? And, most important, who bore the responsibility for providing care? For the LBS, care included not only medicines but also food, heat, linens, and a caregiver. Chronically scarce resources and fluctuating demand for care raised a perplexing and enduring question as to whether care at home was a public, private, or cooperative, community responsibility. The LBS recognized that some patients could not be cared for at home, even with the services the ladies of the society provided.

Charleston in 1800

In 1800, Charleston was the fifth most populous city in the country, with a population of 10,104 blacks (slave and free) and 8,820 whites. The city's development differed remarkably from that of other large cities—Philadelphia, New York, or Boston. Its climate, its evolving agricultural society, and the institutions of slavery and Anglicanism shaped a very distinct worldview.[2] Charlestonians did not cluster by race, class, or function. Merchants, professionals, and planters intermingled within city neighborhoods, and slaves lived in adjacent dependencies. The poor generally lived in crowded alleyways near the wharfs or at the northern edge of the city.[3]

A small number of city residents enjoyed a luxurious life. Family connections, agreeable manners, sociability, and conspicuous leisure were often described as essential ingredients in the lives of Charleston's wealthy class of merchants and planters. Most residents, however, were much less fortunate, with enormous inequities between the rich and the poor and virtually no middle class.[4]

Charleston's prosperity was considerably undermined by the stultifying impact of the War of 1812 and the depression that followed. Jobs were scarce for working-class men and women, and the numbers of poor were steadily increasing. The white working class competed with slaves who were "hired out" by their "masters" for skilled and unskilled jobs, both public and private. Slave labor accounted for 60 percent of the city's workforce and was concentrated largely in domestic service.[5]

Change in the economic conditions of poor blacks and whites remained unlikely as long as Charleston's upper class resisted growth in the city's commercial or industrial establishments. Charleston's history was also influenced by a series of disasters: hurricanes, great fires, earthquakes, and epidemics were regular occurrences, with poverty and disease their frequent companions. The city was most unhealthy, its inhabitants regularly visited by malaria, yellow fever, dysentery, smallpox, dengue fever, diphtheria, typhoid, measles, and tuberculosis. Charleston was described as heaven in the spring, hell in the summer, and a hospital in the fall. The poor were the most common victims of both disease and the effects of disasters on the commercial interests of the city.[6]

South Carolina Medicine before 1860

The practice of medicine and nursing in late nineteenth-century Charleston was typical of the times. Before the founding of the Medical College in 1824, South Carolinians studied medicine in Philadelphia or New York, and many new doctors sought postgraduate training in Europe, especially Paris. Initially, Charleston physicians retained the influence of the renowned physician Benjamin Rush and the medical teachings in Philadelphia, but by mid-century French medical practice was gaining popularity. The new emphasis was on pharmacotherapeutics, namely, the medical use of opium, quinine, calomel, rhubarb, magnesia, castor oil, ipecac, squill, and mercury, supplemented with turpentine, iodide of potassium, and gentian. To fill the demand for these remedies, Charleston had twenty-two apothecary shops; surgery was employed only for traumatic conditions. Charleston's reputation as a rich and unhealthy city ensured that physicians were always in good supply for the wealthy.[7]

Although nurses are mentioned in plantation manuals, much less is known about nurses than about physicians during this period. One document describes the nurse's charge as caring for the sick, fulfilling the orders of the overseer or the doctor, and preparing food for the sick. Nurses also were responsible for the cleanliness of the bedding and utensils. According to the document, care was allegedly the same for slaves and their plantation family owners.[8] As early as 1843, the LBS records mention hiring nurses to care for the sick.[9]

Caretakers of the Sick Poor

In Charleston, laws designating responsibility for the poor date from the 1690s. Later, relief of the "pauper sick" was the responsibility of the Vestry of St. Phillip's Episcopal Church. By 1736, a small hospital had been established to care for the sick poor, but it housed primarily chronic invalids. Most people received medical attention from the parish doctor in their homes or lodgings. After the Revolution, responsibility for the care of the sick poor passed from the church to civil authorities. Charleston's evolving sense of obligation to the poor eventually resulted in a rather distinctive system of benevolence. With privately endowed institutions for orphans and the sick administered by boards appointed by city government, and the elite distributing charity from both public and private sources, benevolence in Charleston fell within the political sphere. Simultaneously, more than thirty private and religious organizations included charity among their many activities, although organizations formed by women generally limited their activities to purely philanthropic concerns.[10]

The city's first major expenditure on behalf of the poor was the construction of a three-story brick almshouse. The public almshouse provided shelter for the destitute, the sick poor, the insane, women in childbirth, chronic invalids, vagrants, drunkards in need of reform, and deranged paupers. It was also used as a house of correction for slaves, vagrants, fugitive seamen, and disorderly persons, bond or free. No definitive separation was made between the hospital area and the custodial portion of the almshouse, except to place the sick on the second floor. Space for the sick increased or decreased with demand. Filthy and crowded, the almshouse was viewed by the poor as a place of punishment for the unworthy rather than a house of refuge for the destitute. The poor could also apply for "outdoor relief," usually in the form of food rations.[11]

While the Commissioners of the Poor believed inducing the poor to come into the almshouse was more economical, and less susceptible to abuse, than care outside, the large number of women and children receiving relief made institutionalization an impractical undertaking.[12] Not until 1842 could blacks receive rations or be admitted to the poorhouse, unless they were insane. In that year, a special committee was appointed to examine the treatment of "free colored poor." Acknowledging that Charleston's free people of color, who numbered approximately two thousand, had been taxed to support the poorhouse, the Commissioners of the Poor concluded that the dictates of justice and humanity called for making them rightful recipients of public relief. Recognizing that the distinctions of caste must be strictly and broadly preserved in slave-holding communities, the commissioners recommended that free blacks be cared for in a place separate from the white poor.[13]

In addition to care at the almshouse, medical services for the poor in

Charleston were also provided through a privately endowed, publicly supervised dispensary (the Shirras Dispensary). Interest in a public dispensary dated to 1784, but not until the early 1800s did this become a reality. The intended recipients, the "worthy poor," received medications at the dispensary or, if unable to get to the dispensary, were visited in their homes by one of the dispensary physicians. The city also maintained a quarantine (pesthouse) hospital and a marine hospital and assembled temporary facilities when epidemics swept the city. Additionally, several small privately run hospitals for slaves provided medical attendance, medicines, board, nursing, and, for an extra fee, surgery.[14]

"Time Stood Still": The Ladies Benevolent Society's First Fifty Years

For anyone who did not fall within the purview of the almshouse, dispensary, or slave hospitals and was unable to afford physician visits or hired nurses, the LBS offered some hope for relief by administering to the sick at home. Founded in 1813, during the British blockade of Charleston harbor, the LBS was a response to the poverty produced by the War of 1812, as well as the devastating effects on the poor of Charleston's frequent epidemics. The society was organized in September 1813, in the city's most unhealthy season, by 125 women—the wives, sisters, and daughters of Charleston's wealthiest families.[15]

In this slave-owning society, in which the primary responsibility for at least half the population from cradle to grave rested with "masters," benevolence for the remainder of the population was personal and immediate. As the city's mayor explained in 1835, men should be kind to each other, because man is essentially a dependent creature. The whole social system is but a "chain of reciprocal dependence, the poor hanging upon the rich, and the rich upon the poor." The rich were obligated to improve the conditions of and provide for the comfort of the poor, and the poor were expected to manifest their gratitude to the rich.[16]

Like the exclusively male societies doing charity work, the LBS formed an organization that depended on a perceived intimate understanding of the poor and sense of responsibility toward them. But, unlike its male counterparts, the LBS was not a mutual benefit association and did not provide social activities for its members. The society was solely philanthropic in its purpose. The historians Jane and William Pease have suggested that members of the LBS, unlike rich men, were restricted to activities appropriate to their sex and in "the meek and quiet spirit, the humility and the desire to profit by the counsel of the wise and good."[17] Whether the LBS's narrowed focus came from cultural restrictions, as Pease and Pease suggest, or from a desire for more meaningful outlets for members' leisure time is difficult to know.

In 1813, the LBS was Charleston's only purely charitable private society. It added to the available medical services for Charlestonians with few avenues of care, namely, free blacks and poor whites. The LBS provided an estimated 10 percent of the benevolent dollars spent in Charleston for the sick and needy and thus had an ample agenda. The ladies not only raised needed funds for care of the sick and poor, they also visited homes to distribute their largesse.[18]

The LBS was a religious as well as a benevolent institution. "Ministering unto Him who for our sakes became poor" was for these ladies the highest of Christian privileges. Entering the "obscure hovels of sickness and death, doing that work which God has assigned us," they sought to "discountenance vice and encourage virtue."[19]

By December 1814, the ladies of the LBS had incorporated, established rules to guide their activities, and accumulated sufficient funds to care for some of the city's sick poor. During these early years, direction was provided by a superintendent, a junior superintendent, a treasurer/secretary, and twelve board members. The constitution stipulated that the treasurer must be an unmarried lady, reflecting the society's desire to retain complete control of its property (at that time, a woman's assets were transferred to her husband on marriage).[20]

A visiting committee conducted the daily work of the LBS. As would be expected, patient load varied with the seasons and the occurrence of epidemics. In the early years, the ladies cared for an average of 290 patients annually.[21] Lady visitors served several important functions, including ensuring the LBS did not sanction or encourage idleness or vice. The proper objects of the society's work were clearly those poor "whose strength was gone, when consuming disease has laid them low, and bodily pains are rendered more acute by the sad thought that their families are suffering from their inability to labor for them." Persons of loose or disorderly conduct, whose "sickness sprang from intemperance," who had no families or were not under the particular care of some relative or friend, or who were destitute of the means of support, whether in sickness or in health, were considered the proper inmates of the almshouse. By 1844, a list of persons unworthy of assistance was created to guide each lady visitor.[22]

The ladies visited every home that could be entered with propriety. Certain streets and alleys were never visited, nor were persons of suspicious character; under these circumstances, the ladies immediately sent relief or the dispensary physician. The ladies nursed the sick and provided them with the simplest comforts: sugar, coffee, lamp oil, grits, wood, barley, soap, and meat or chicken if ordered by the physician. Flannel and a blanket were provided when necessary, as well as clean beds, sheets, and pavilions (tents) in summer. The visitors were allowed discretionary powers to prescribe the exact quantity and quality of supplies for patients with various types of disease. In difficult or unusual cases, ladies of the visiting committee (especially its youthful members) sought the advice and instruction of the board or superintendent.[23]

The ladies' intent was to judiciously create an individualized, intimate, and ordered method of caregiving for the destitute that interlocked with the city's other benevolent organizations. Visitors were expected to ask patients whether they had availed themselves of assistance from other institutions and, if they had not, to assist them in obtaining church money, rations from the poorhouse, or other help. Medicines were obtained only from the dispensary and no others were to be bought or paid for by the visitors, since the dispensary physician was seen as the most competent judge of the patient's constitution and disease. Dispensary physicians were given the names of the visitors and referred to them all persons worthy of relief.[24] Most important, the LBS supplied the sick poor with nurses, for "of what avail are medicines or proper nourishment, unless there be some kind hand to administer them in due season?" Hired nurses (untrained) were paid according to the services required. A relative or friend with whom the patient resided, if competent and trustworthy, was often thought preferable to a stranger and was paid as the nurse.[25]

The "Vexed Question of the Chronic Patients"

Although LBS rules guided the visitors, these rules did little to rationalize decisions about the single most persistent problem, what Superintendent Rebecca Holmes later called the "vexed question of the chronic patients." Although the rules were quite clear about the proper treatment of those with chronic disease—they were the responsibility of the almshouse—debate over their proper disposition resulted in the first documented differences of opinion about the proper mission of the society.[26]

The ladies occasionally found it difficult to stop providing assistance to the worthy poor who were no longer acutely ill, and for years they flexibly interpreted the rules. By 1825, persons "permanently on the bounty of the Society" included seven women—four whites and three free blacks. That summer, a committee of three ladies visited each pensioner. Their report described why these pensioners needed to receive charity at home rather than being sent to the almshouse. All lived in the most desperate situations and most were elderly. They were described as respectable women who had been hard working and had raised and educated their children, having supported themselves variously as a schoolmistress, a nurse, a shopkeeper, or through needlework. Now they were weak, crippled, and suffering from a variety of illnesses from dropsy to leprosy. Most pitiful was one white woman: "Mrs. Cowie is an old woman—she has leprosy—and so long, and so greatly has she suffered under it, that her hands are drawn up and deformed. Her eyes are in a state of inflammation—and her body a perfect Skeleton. She is indeed a pitiable object—calling forth compassion of every beholder . . . our hearts bled when we beheld her."[27]

The committee found that poor white women were better off than free black women, because white women could receive rations from the almshouse and more often obtained support from their churches. Free black women, who at that time (1825) could not receive outdoor relief or be admitted to the almshouse unless insane, had few avenues for assistance. Some free black pensioners on the rolls of the LBS received a "pittance" or occasionally had their rent paid by their church. Forced to depend on the LBS, they grew old under the ladies' care. Having carefully examined in detail the circumstances of each woman, the committee found the needs of these pensioners to be great.[28]

The number of regular pensioners cared for by the LBS gradually increased, reaching twenty by 1840. As the number of persons needing permanent care by the society grew, visitors found themselves under greater pressure to discharge patients from their care before they crossed the line between acute and chronic illness. Visitors were asked not to countenance patients longer than necessary; visits and supplies were to be continued only one week after the patient was discharged by the physician; and when possible visitors were changed each quarter. Keeping patients longer than necessary reduced the society's funds and destroyed the object of its founders.

Frugality and avoidance of too much closeness were recommended. Those with chronic disease were to be assisted to the almshouse. If they refused to go, "They must be left to do as they please—but they are not to be continued on the Society."[29]

New rules limited to eight the number of aged and infirmed persons permanently cared for by the LBS. Pensioner status was granted only after careful investigation of character and circumstances and on recommendation by a member of the visiting committee. Pensioner status required an affirmative vote of two-thirds of the members and approval of the superintendent.

The Invention of "Medically Necessary Care"

Redefinition of the LBS's role and commitment to the chronically ill changed its relationship with the dispensary physicians. Previously, the society had sent the names of its visitors to attending physicians, who, each quarter, had informed the ladies of all persons worthy of relief. The relationship would now be both more formal and more professional. Anyone seeking the aid of the LBS must first apply to the Shirras Dispensary for medical care and obtain certification of sickness. By asking for the physician's certification, the LBS hoped to "relieve the visitors of the embarrassment frequently felt to know whether or not the applicants for aid were really sick." Furthermore, the dispensary physicians helped the lady visitors proceed in the spirit of the society by recertifying each patient's illness every week. Thus, as early as 1850, medical necessity and recertification of need were established as a technique for "rationalizing" the selection of appropriate recipients of care.[30]

Despite the new rules and strident rhetoric, the ladies remained incapable of taking a firm stand. While the number of chronically ill pensioners could not exceed the prescribed number of eight, the visitors nevertheless failed to discharge patients quickly. By 1857, Superintendent Rebecca Holmes reluctantly admitted that she found herself embarrassed by patients who, from quarter to quarter, were assisted without much hope of recovery. She agreed that discharging patients did "violence to their [visitors'] benevolent feelings" but asserted that they must all learn to discharge patients who were "not quite well."[31]

The plight of the chronically ill vividly illustrates the complex relationship between mission and economics. Though the chronically ill seemed reasonable recipients of the ladies' largesse, they were never part of the stated mission of the LBS. The fate of chronically ill patients was quietly determined by the views of the membership, which clearly fluctuated with the economic fortunes of the LBS. Many members of the society acknowledged that the poorhouse was a choice of last resort, but periodic forays into the care of the chronically ill served as a reminder of the wisdom of the society's original mission. It was difficult to choose between the more and the less deserving. However, limitations on, or even outright exclusion of, the chronically ill avoided the rapid depletion of funds and possible destruction of the LBS altogether. Should more money be raised, these difficult choices could be avoided.

Free Persons of Color and the Question of Race

By the mid-1800s, 3,441 "free persons of color" lived in Charleston.[32] All free blacks suffered from the oppressive burden of racism, but there was a comparatively prosperous and cultured mulatto elite that assumed more authority in the churches than did the slaves. The beneficent concerns of this small group led to the formation of several mutual benefit and charitable societies. Most free blacks, however, did not enjoy a significant place in Charleston's economic order, and they struggled for their well-being. The needs of poor free blacks often exceeded the resources of the black community. The LBS was one of the few white benevolent organizations willing to provide assistance to free blacks during illness.

The combination of growing numbers of free blacks, chronic illnesses among the poor (whether black or white), and limited funds presented an especially complex problem for the LBS. Unreported in the minutes of society meetings, but obviously a reality, was the conflict between caring for free blacks and charity for needy and deserving poor whites. Dependent on the continuing support of its members, the leadership of the LBS navigated Charleston's complex race rules with care.[33] The ladies' first recorded debate about issues of race occurred in 1825. The record reports only the rereading of their original "resolution respecting the free people of color"; the ladies temporarily agreed that their mission to relieve the distresses of the poor and administer comfort to the sick of all races would be continued.[34] Reflecting concerns about the propriety of their continued assistance to free persons of color in the face of South Carolina's increasingly aggressive efforts to close all avenues to emancipation, the ladies would later debate and seek legal advice on the ever-changing meaning of *free*. Resorting to the great southern tradition of "innocent because we did not ask," the ladies concluded that they did not need to inquire how freedom was obtained. As long as the person was de facto free—that is, "not held or used as a slave by any owner"—the ladies could provide care without fear of criticism.[35]

The dilemma of what to do with free blacks who were chronically ill was solved, at least in the short term, with a bequest of $1,300 from John M. Hopkins, a Quaker businessman. Hopkins established this trust to care for Marie Drayton, "a poor old and infirmed free woman of color residing in Charleston during the rest of her life," with the remaining income to "relieve the free sick and infirmed persons of color" of the city. Suddenly in a unique position to aid those "peculiarly" destitute, and with a deep sense of the complex issues involved in accepting this trust, the LBS board of managers deemed it necessary to consult a gentleman ("trust man") whose judgment and expertise would guide this new duty.[36]

To "guard against imposition," the ladies appointed an oversight committee for the Hopkins fund. At the recommendation of the "trust man" and following the example of the almshouse, applicants for pensions were required to bring a certificate signed by a "respectable white person" describing their "character and necessities." At the discretion of the committee, pensioners were assisted with rent and furnished with food or clothing. Ever anxious to carefully discharge its new duties, the board of managers kept separate books, documented any assistance given, and recorded the names of pensioners and their sponsors. The committee was also charged to prepare an annual report, noting the number of persons cared for and the general effects of the charity.[37]

Whatever the intentions of the fund's benefactor, the numbers of Hopkins pensioners, along with other pensioners supported by the LBS, continued to creep upward. Despite a decision in 1842 to give free blacks access to outdoor rations and the almshouse, the number of Hopkins pensioners had reached fifty by 1847.[38] In 1855, the Hopkins fund increased by $2,133 with a bequest from Mrs. Elizabeth Kohne, a member of the LBS since 1817.[39]

A "Trifle in the Treasury"

From its founding in 1813, the Ladies Benevolent Society was dependent on private funds from annual subscriptions ($5), life subscriptions ($50), donations, legacies, and interest on stocks and bank shares. The society's financial security and its capacity to extend its usefulness fluctuated from year to year with the generosity of contributors, illness and need in the city, and the success of investments. The ladies constantly hoped that subscriptions and dividends would equal or exceed expenditures, but this rarely happened; most years began with a "trifle in the treasury." Each year brought losses in membership through relocation, resignation, and death. Membership declined from a high of 346 in 1824 to a low of 162 just prior to the Civil War. Only in years of terrifying epidemics was this downward trend in membership reversed, and then only temporarily.[40]

Sometimes the LBS was blessed with substantial donations and bequests.[41] Relishing those rare occasions when they could proceed without "anxiety as to the competency" of their funds, the ladies nevertheless realized they could not rely on such donations and the only real remedy was to increase the number of subscribers.

By the eve of the Civil War, the ladies were unable to extend the society's usefulness. For the LBS time had stood still. While it cared for an average of 184 patients each year and raised an annual budget of about $3,000, membership was declining.[42] Never able to replicate the "zeal" of earlier superintendents, the ladies simply persisted in their traditional mission.

By midcentury, the LBS was no longer the only purely charitable enterprise in town. The Sisters of Charity of Our Lady of Mercy, the Methodist Benevolent Society, the Howard Association, and the Young Men's Christian Association were also caring for the city's sick and poor. The sick poor could also turn to the newly established Roper Hospital. Anxious to extend limited funds as far as possible and viewing any duplication of services as "a mistake and imposition," the LBS created a variety of cooperative arrangements with other organizations, especially during epidemics. Anticipating future trends, the ladies acknowledged that since establishment of "that excellent institution" the Roper Hospital, the list of patients requesting their services had declined. By 1860, the LBS's future seemed particularly difficult to predict.[43]

Nevertheless, the ladies of Charleston had developed a system of caring that was receptive to the needs of the poor yet respectful of the city's particular way of life. Building on their domestic expertise and concepts of civil and religious duty, they had created a highly individualized, intimate, and ordered method of caring for the destitute that carefully meshed with the efforts of the city's other benevolent organizations. As the direct distributors of charity, they had evolved a code of ethics that maintained social distance even when physical distance diminished. Despite fifty years of experimentation, however, the formula for efficient yet sensitive rationing of scarce resources remained elusive. What had begun as individual uplift and good works associated with their faith became a mission complicated by the unpredictable variables of caring for the sick at home—family circumstances, chronic disease, and poverty.

Part I

Inventing Home Care in the
Nineteenth Century

CHAPTER 1

Trained Nurses for the Sick Poor

*A*s Charleston struggled to recover from the Civil War and Reconstruction, many other U.S. cities were experiencing disparate but equally challenging transformations. Most dramatically changed by the end of the nineteenth century were northern coastal cities with large concentrations of immigrants and industry, a combination making poverty, filth, and disease unavoidable. Population growth alone required major adjustments in the lives of many city dwellers, even those able to escape to newer, less crowded neighborhoods.

With immigration accounting for much of this growth, ethnic, cultural, religious, and economic differences accentuated the separation between old and new inhabitants. The shattered remains of a once cohesive city life and the resulting distance between the classes caused apprehension for many people. Wealthy northerners saw immigrants as a profound threat to the social and moral order. The search for solutions to these urban problems included various strategies to stave off perceived moral, social, and physical disintegration among the urban masses. Convictions about the causes and cures of poverty gave rise to scores of charitable organizations. Many such organizations would send trained nurses into the homes of the sick.[1]

In the North, ladies of means not only introduced a novel way of providing care in the home but also transformed voluntarism from a religious impulse to a tough theory of "scientific charity."[2] Fundamental to the idea of scientific charity was a belief that urban poverty was caused by the moral deficiencies and character flaws of the poor. The circumstances of city life removed the poor from the elevating influences of their moral betters and thus, so the argument went, the poor were left without the means of recognizing and correcting their deficiencies. Irresistibly drawn by these ideas, thousands of upper- and middle-class ladies volunteered to expand the moral conduit between the classes by visiting the homes of the poor, thus redeeming American society through "true friendship." These visits also provided the necessary data on which the future laws of charity and reform would be based.[3]

Unfortunately, the friendly visitor, imbued with the spirit of scientific charity, was nevertheless defenseless in the face of influenza, pneumonia, typhus, typhoid fever, summer fever, smallpox, scarlet fever, measles, whooping cough, dysentery, and tuberculosis. Despite declining death rates, late-nineteenth-century U.S. cities were unhealthy places to live (Figure 1). It became increasingly difficult to deny the relationship between disease and poverty.[4]

Motivated by a shared vision of the good society, wealthy ladies in New York, Philadelphia, Boston, Buffalo, and Chicago began to hire trained nurses to "bring care, cleanliness, and character" to the homes of the sick poor. At first they were unaware of any similar activities in other cities, much less the earlier work of the Ladies Benevolent Society in Charleston. Many modeled their new charities on district nursing in England, and they visited or corresponded with its founders.[5]

The English Model

The English model that so attracted lady philanthropists in the United States evolved in response to "the rising tide of pauperism" in England. The first attempt to send trained nurses into the homes of the sick poor was initiated in Liverpool in 1859 under the direction of William Rathbone, a wealthy Quaker philanthropist and businessman.[6]

Rathbone first encountered a trained nurse during the long and painful illness of his wife. Amazed by the nurse's ability to bring comfort and ease in luxurious circumstances, Rathbone wondered what she might do to alleviate suffering in the more wretched homes of the poor. He therefore persuaded his wife's nurse, Mary Robinson, to participate in an experiment: to care for the sick poor while simultaneously teaching them to better care for themselves.[7]

Initially, Mary Robinson was shocked and overwhelmed by the hopeless misery of poverty and reportedly tried to terminate the experiment before

FIGURE 1. Philadelphia Street Scene, 1886.
Visiting Nurse Society of Philadelphia, Center for the Study of the History of Nursing, School of Nursing, University of Pennsylvania.

the end of her three-month agreement with Rathbone. But encouraged by her employer, who believed in constructive charitable intervention, Robinson continued, and eventually both Robinson and Rathbone declared the project a great success. In the words of Rathbone, the hopeless were restored to health, breadwinners and mothers were restored to independence, and the spread of weakness and disease was halted. Nursing had, he declared, demonstrated its ability to prevent the "moral ruin, the recklessness, the drunkenness, and the crime which so often follow upon hopeless misery."[8]

Having witnessed such success, Rathbone saw it as his duty to extend the benefits of trained nursing care to all of Liverpool's poor. Support for this enterprise far exceeded the available supply of trained nurses, and by 1861 Rathbone turned in desperation to Florence Nightingale.[9]

Nightingale viewed care of the sick poor in their homes as one of nursing's most important tasks and willingly lent support to the new enterprise. A voluminous correspondence between the two ensued, and Nightingale closely scrutinized and considered all Rathbone's plans. Her earliest ideas on the subject of nursing the sick poor appeared in 1865, in the Committee of Home and Training School's *Organization of Nursing in a Large Town*. But it was Nightingale's 1874 paper "Suggestions for Improving the Nursing Service of Hospitals and on Methods of Training Nurses for the Sick Poor" that most influenced development of district nursing and guided every organization promoting this model of care.[10]

Two years after writing this paper, in a widely read newspaper article, "Untrained Nursing for the Sick Poor," Nightingale first introduced the public to the work of the district nurse. Her plea for financial support was bolstered by her description of the "depauperizing" effects of trained nurses on the lives of the sick poor. These visitors to the homes of poor were not, she assured her readers, some new form of cooks, relief officers, district visitors, letter writers, general store keepers, upholsterers, almoners, purveyors, ladies bountiful, head dispensers, or a medical comforts shop; they were simply nurses. Their laudable results were achieved through nursing care and by scouring dens of foulness, dirt, and vermin, turning them into the tidy, airy rooms required for recovery. If other forms of relief were required, the nurse knew how and where to apply for them, but her real task was simple and straightforward: "To set these poor sick people going again . . . with a sound and clean house, as well as with a sound body and mind." It was, wrote Nightingale, "as great a benefit as we can give them—worth acres of gifts and relief."[11]

Although Nightingale offered advice and direction, she could not find Rathbone any trained nurses. Rathbone concluded that he needed to form his own school of nursing, and within four years the school made trained nurses available throughout Liverpool. He organized his nursing service in the same way that he approached the giving of relief: on a district basis under the supervision of ladies. Thus, nursing of the sick poor in their homes came to be known as *district nursing*. As in other forms of relief giving, the presence of ladies guaranteed that the work of the nurses would be supported by appropriate patient records, financial accounts, and, most important, money.

Thanks to the success of district nursing in Liverpool, care of the sick poor grew in popularity throughout England. Somewhat alarmingly to some, however, it gradually became an activity pursued by both trained and untrained nurses. By 1874, a committee was appointed, chaired by Rathbone, to consider whether more adequately trained nurses were needed for this work. Florence Lee, one of the first pupils of the Nightingale School, conducted the investigation, a yearlong study financed by Rathbone—the first of many such inquiries.[12]

Lee, like her mentor Nightingale, went in search of facts and statistics. She wanted to know about the work district nurses did, their training for this work, and the quality and amount of service they provided. She found that most district nurses were taken from "the lower grades of women," that their work generally lacked "regulation" and, as a result, was often open to "grave objections." Most nurses with whom she visited were not, in her estimation, sufficiently trained to care for the sick in their homes. They tended to lapse into slovenliness and neglect, nursing too little while giving too much relief; they communicated infrequently with doctors and often acted, in the absence of medical supervision, more like a doctor than a nurse. Finally, many of these district nurses failed to instruct the patient's family and friends in care of the sick or the necessary preventive and sanitary measures.

Lee found the responsibilities of the district nurse much greater than those of the carefully supervised private-duty nurse, who had the benefits of a doctor close at hand and the "appliances" necessary for patient care. An additional three to six months' training in district work seemed an immediate solution, but Lee concluded that future district nurses should be drawn from a different class of women and should receive more education. Reforming and recreating the homes of the poor were difficult tasks for nurses who came from the same class as their patients. Gentlewomen, she concluded, were most suited for this work, since they could call upon the influence provided by their higher social position. Reciprocally, of all employment open to gentlewomen, none, submitted Lee, was more suitable than district nursing.

Lee's final recommendation was that district nurses be provided with a common home, one that supported them morally, materially, and spiritually, for no woman should be expected to come home "dog tired" from her patients and have to cook and clean. To create a home for the poor, she argued, the nurse must have a fit home for herself. Otherwise, it was like expecting the homes of the poor to be reconstructed by women who had forgotten what a home was. Finally, the head of the nurses' home must be an eminently trained and skilled nurse who would provide the nurses in her charge with training, supervision, support, and sympathy in their common work. In short, what was required was a home for nurses where any good mother, of whatever class, would be willing to let her daughter, however attractive or highly educated, reside.

The committee's report, based on Florence Lee's investigations, was published on 11 June 1875. It was widely read and in such demand that it was soon reprinted. While the committee and Florence Nightingale wholeheartedly supported most of Lee's conclusions, they viewed her endorsement of the gentlewoman as a savior of district nursing with some skepticism. They agreed, nevertheless, to give her methods a try for a "year or so." Lee was asked to implement her plans as Superintendent General of the Metropolitan and National Nursing Association for Providing Trained Nurses for the Sick Poor, created the following winter in London. Ultimately, the committee claimed that its decision to support Lee's recommendations proved correct. Her work resulted in the creation of one grand system, with district nurses' work throughout England under the control of a standardized national organization.[13] The organization Lee had established in England by 1889 remained the standard emulated by American nursing leaders for the next three decades.[14]

The Theory and Practice of Visiting Nursing in America

Guided by English precedent, the introduction of district nursing in the United States was a logical extension of the religion-dominated philanthropic activities of the preceding generation. Florence Nightingale's popular allure, combined with the growing availability of trained nurses, quickly produced a "new power for good" among those fighting the bodily, moral, and social disintegration associated with industrialization, urbanization, and immigration. By the end of the century, what had begun as a self-conscious copy of the English system was evolving into a specifically American approach to home-based care, one characterized by individualism and pluralism. England's district nurse became America's visiting nurse.

American ladies moved beyond the English model's dependence on environmental interpretations of disease to include the germ theory. Despite Nightingale's resistance to the germ theory, Americans increasingly recognized infectious diseases as a greater threat to the public's health than any defects in the character or ambitions of the poor. The knowledge that the diseases of workers who sewed clothes in filthy tenement homes or who processed food could spread to decent, clean, and respectable families was a powerful incentive to eliminate the menace of illness among the poor, particularly immigrants. These were the classes whose "poverty, ignorance, and lack of moral strength entail terrible evils, they are often helpless and even where they try to assist each other, their absolute lack of knowledge of the laws of health may lead to more harm than good from their well-meant endeavor." Trained nurses were both a symbolic and a practical solution to a most intolerable problem—the "dangerously ill"[15] Helen Jenks, founder of the Visiting Nurse Society of Philadelphia, reminded her supporters in 1886 that they could not expect freedom from disease in their own homes unless all classes were taught the laws of health and prevention of contagion.[16] No longer would the activities of the poor during illness be unregulated. No longer would the poor be left to struggle with illness as best they could in their crowded and dirty homes. Nurses would bring a message with their medicines. As disciplined and well-bred women, they would raise the level of household life with their delicate instructions and firm convictions and would protect the public from the spread of disease with forceful yet tactful lessons in physical and moral hygiene. In the trained nurse was found "the safest and most practical means of bridging the gulf which lies between the classes and masses."[17]

Because the work of the visiting nurse would clearly continue to grow, a major concern for those already in the field was how to find an adequate supply of the right kind of nurse. Because the visiting nurse often found herself working alone, the most basic requirement was training in a large general hospital that provided a wide range of clinical experiences. While almost any woman could learn the necessary "theories, systems, and methods," few possessed the other essential qualifications required for this type of work: professional knowledge and "womanly training." The crucial characteristic that no amount of hospital work could cultivate was the natural gift of personality, which in the case of the visiting nurse meant tact, gentleness, sympathy, ingenuity, resourcefulness, and firmness. In visiting nursing, claimed one authority, more than in any other branch of the profession, a nurse should be born, not made—but women of this higher caliber were in short supply.[18]

Visiting nursing involved much more than good bedside nursing and extended well beyond caring for patients in clean beds or an orderly ward. Disregarding the claims of critics that the sick poor "would be content if their mere physical needs [were] attended," the lady managers believed a "higher-class" system was required. They argued that the professional authority of the nurse, combined with the dependent condition of the patient, created the most efficient way of bringing into the homes of the poor the "rudimentary principles of the laws of physical wellbeing."[19] Nurses were expected to treat each poor person with the same courtesy, gentleness, and consideration they would use in the homes of the more fortunate.[20] As one nurse explained, on entering the patient's home, nurses understood that the patient and family circumstances were as important as diet or treatment. While the doctor diagnosed disease, the nurse diagnosed the whole situation. The nurse had to convince others over whom she had no authority to follow her directions. At this point, personality made all the difference in outcome. If successful, the good visiting nurse could make do with the meager resources found in the home and create an environment conducive to the patient's recovery, while simultaneously coaxing or reasoning the family into maintaining this environment in her absence.[21]

Gaining the friendship and confidence of families in her care was only the beginning of the visiting nurse's job. She was also expected to go into the community, make herself known to other families, and thus find patients rather than simply wait for patients to find her. Finally, the nurse needed the right personality to meet her patients' complex medical and social needs and to work in harmony with other charitable organizations, as well as the doctors in her district.[22] As nurse-author Mary Gardner suggested, the ideal visiting nurse was a faultless creature "possessing all the virtues, combining the experience of age with the enthusiasm of youth and also having a sense of humor, which is perhaps the only thing which will make the years" of this kind of work possible. For this, nurses in 1909 earned between $600 and

$900 a year. Few were paid a "fitting salary," and salary—or what later came to be called "a living wage"—would remain a point of chronic tension between the nurses and their lady employers.[23] Unlike their colleagues in private-duty nursing, who earned more, visiting nurses could at least count on steady employment.[24]

Many nurses, while attracted to visiting nursing, found the work too mentally and physically exhausting. Walking long distances in all kinds of weather, climbing endless flights of stairs, cleaning and disinfecting patients' rooms, changing beds, and being constantly exposed to disease were all part of the visiting nurse's daily routine. The "delicate" nurse found this an impossible undertaking, but even the strongest became exhausted—even sick—at the end of a day of work. Some were simply too revolted by the filth, vermin, and dirt typical of overcrowded tenements. Fatigued, discouraged, and often sick, many nurses left for more lucrative or easier work, while others left to marry or were called home because of death or illness in the family. As a result, the turnover was high and replacements difficult to find. With a large proportion of the staff leaving, each year seemed a new enterprise. To help nurses adapt to the challenge of this work, some of the larger visiting nurse associations (VNAs) established postgraduate courses. The first was begun in Boston in 1906; by 1920, some twenty programs were operating around the country.[25]

Visiting nursing appealed both to U.S. philanthropists seeking greater order in the city and to a newly emerging nursing leadership striving for recognition as a profession. Nursing leaders such as Isabel Hampton recognized in this new kind of work the potential for fulfillment of the womanly mission of social uplift and enhancement of the status of nursing. Visiting nurses could attain a distinct social importance as part of "the highest type of nursing." Unlike their hospital-based sisters, these nurses working alone in the homes of the poor occupied center stage.[26] Hampton's views on visiting nursing began to crystallize after her visit with the ladies of the Boston Instructive District Nursing Association in the spring of 1890, and in the summer of that year she observed the English system during a visit to London. In 1893, as organizer of the nursing sessions of the International Congress of Charities, Corrections, and Philanthropy at the Chicago World's Fair, Hampton had the opportunity to further shape the future of nursing. Three of the four speakers on district nursing were from England, reflecting the authority and influence of English nursing. For the first time, American visiting nurses and their lady supporters had the opportunity to discuss at length with their English colleagues the pressing issues they shared. The papers and subsequent discussions at this meeting no doubt strongly influenced later developments in the United States.[27]

The rising cost of hospital care and a growing willingness to support any effort to reduce the number of charity patients seeking hospitalization further contributed to rapid acceptance of visiting nursing. In an effort to shed their stigma as a social service for the needy, hospitals valued any program that made them more attractive to patients able to pay for care. Obviously, one way to relieve the financial burden of hospitals—asserted by some to be caused by excessive and indiscriminate charity—was to provide the poor with more care at home while simultaneously teaching them to stay healthy.[28]

In recalling her 1905 "grand tour" of the country's leading VNAs, Mary Gardner described the diversity of aims and methods. Earnest and eager, each organization had acquired the habit of original thinking, "working as if on a desert island remote from any shore."[29] In Gardner's estimation, these organizations were simply "blundering along . . . acting out of total ignorance," but their individualistic behavior reflected the pluralistic nature of U.S. health care. A 1909 survey of the 565 organizations sponsoring the work of visiting nurses largely confirmed Gardner's observations. Most striking were the multiplicity of and changes in sponsorships, the variation from community to community, and the dominance of small, potentially vulnerable organizations.[30]

From a contemporary perspective, one might have expected VNAs, boards of health and education, and hospitals to sponsor home-based nursing care. The list of 1909 sponsors, however, was somewhat curious. Most were "voluntary," philanthropically financed organizations that no longer exist or, if they do still exist, rarely have health-related missions. Organizations hiring visiting nurses before and after the turn of the century exemplified both changing sources of sponsorship and a new conceptualization of the needs of the sick poor. They reflected an evolving system of social and medical care in which "scientific" explanations and interventions would replace those of the church-based urban philanthropy that dominated the preceding generation.[31]

Before 1900, the major sponsors of visiting nurses were VNAs (24%), hospitals (15%), and church groups (11%). After the turn of the century, such sponsorships began to decline, especially those by hospitals and church groups. For hospitals, withdrawal from home-based care embodied changing agendas and economies. Hospitals had at one time been interested in generating revenues by charging for the services of student nurses sent into private homes. But as hospitals became more successful in sustaining themselves through inpatient care, they increasingly needed student nurses to care for the growing numbers of paying patients. Hospitals' interest in home-based care would not reappear for another fifty years.[32] By 1909, church sponsorship of home-based care had also decreased dramatically. Mission Boards, King's Daughters (a religious and philanthropic women's group), dispensaries, and settlement houses also experienced a slight decline, suggesting all were becoming marginal to an emerging system of medical care. Although Charity Organization Societies had not yet experienced comparable changes, they would eventually end their sponsorship of home-based nursing care.

Even though by 1909 the numbers of VNAs had declined slightly, they remained the cornerstone of care for the sick poor at home. With the combination of the philanthropic concern of the lady managers and the scientific knowledge of the nurses, VNAs dominated the development of visiting nursing in the years to come.[33]

Changes in sponsorship for home-based care were also attributable to rapid growth in public, tax-supported boards of health. As "foot soldiers" in the modern campaign for public health, trained nurses were sent into the homes of the poor to teach the ways of healthful living. Concepts of disease, personal responsibility for health, and the latest advancements in medical science and public health formed the nurse's message in these new programs.[34]

Growth in preventive services varied from city to city, with voluntary and governmental agencies assuming unpredictable and often overlapping responsibilities. As the confusion grew, so did the debate about the shared and differing functions of voluntary and governmental health agencies. As a former health commissioner of New York City later recalled, "Competition and rivalry in methods, resources, and accomplishments became as keen as selling soap or advertising toothpaste."[35]

Debate over the proper domain of public health practice was not confined to the struggles between voluntary and governmental organizations. Attempts by health departments to extend the focus of their concerns resulted in conflict with the medical profession. Most health departments were forced to abandon claims to any curative activities construed as threatening the economic well-being of private physicians. Consequently, despite much ongoing discussion, tax-supported health department services became increasingly preventive in scope. An unexpected consequence of this development was the limiting of publicly funded nursing activities to prevention of disease, with home care of the sick left to voluntary agencies, primarily VNAs.[36]

In many larger U.S. cities, the result of this idiosyncratic mix of governmental and voluntary initiatives was a perplexing assortment of uncoordinated nursing services, each driven by its own mission and definition of community need. By 1909, the most extreme example of the growing popular appeal of visiting nurses could be found in New York City, where fifty-eight different organizations sent 372 nurses out into the community. Although visiting nurses could be found across the country, they remained predominantly a northeastern urban phenomenon.[37] Regional variations in the type of organizations sponsoring visiting nurses did exist, however. No single type of agency could be found in all regions and no region had all types of organizations; eleven states had no visiting nurses.[38]

Clearly, the image of the nurse climbing tenement stairs to save the indigent from illness and bad habits had struck the fancy of a wide variety of turn-of-the-century reformers. Although enthusiasm developed slowly at first, the nurses and their lady supporters were confident they had found the best means "at the smallest cost" for helping the "conditions of the poor, sick or well." With growing numbers of immigrants, the ever-present danger of contagious diseases, and the evolving system of modern hospital care, others began to share their views. Visiting nurses rapidly acquired symbolic and practical appeal. The visiting nurse was an irresistible answer to the complex social burdens caused by the illnesses of the urban poor. For nurses, visiting nursing was a new field of employment, offering economic security and professional independence.[39]

CHAPTER 2

Creating Their Own Domain

Ladies, Nurses, and the Sick Poor

O n a cold winter morning in 1902, Mrs. John Lowman, the wife of a prominent physician and a member of the board of trustees of the Visiting Nurse Association of Cleveland, agreed to accompany one of the association's nurses on her rounds. "I shall always remember my surprise," she recalled ten years later, "at the situation we encountered in the first home . . . I sat in a rocking chair, feeling very low in spirit and quite benumbed by astonishment and a kind of terror."[1]

In contrast the nurse, who was young, attractive, neat, and dainty in appearance, seemed to Mrs. Lowman quite at home. While Mrs. Lowman sat, presumably trying to overcome her shock, the nurse attacked the situation with cheerfulness and decision, rapidly setting the sick room and sick patient in order, changing the bed, washing the patient and combing her hair, and reordering the room for proper ventilation and light. She determined the patient's condition and left notes for the physician when he returned. Finally, the "gossipy old woman" neighbor was replaced by a "sensible looking" one who promised to carry out the nurse's instructions.[2] Mrs. Lowman accompanied the nurse on two more visits before reaching the end of her endurance. Finding herself tired, cold, and a little frightened by the scenes into which her young guide had led her, she returned home. Unlike Mrs. Lowman, the nurse seemed entirely competent to handle these terrible situations and proceeded on her way to visit the six or seven patients remaining on her list.[3]

Mrs. Lowman's account of her introduction to visiting nursing nicely illustrates the distinctive yet potentially unstable nature of the functions of lady managers and visiting nurses prior to World War I. As this story suggests, differences between these two groups were not confined simply to areas of professional competence but reflected social class as well. Nurses were not ladies and ladies were not nurses, thus their mutual dependence seemed obvious. Because of their unique training, visiting nurses were a most practical means of caring for the sick poor at home. The lady managers, on the other hand, were singularly qualified to find and manage the money needed to pay the nurses.

This symbiotic relationship remained stable until 1909 or 1910, when, according to prominent visiting nurse Mary Gardner, "popularity struck." With it came a growing demand for visiting nurses not only to care for the sick but to do preventive work with babies, mothers, schoolchildren, and patients with tuberculosis.[4] In addition, after 1909, when the Metropolitan Life Insurance Company began paying for nursing visits for its sick policyholders, visiting nurse associations also provided more services to the working class.

Lady managers, eager to help launch these new fields, found their enthusiasm exceeded their ability to finance and manage agency growth. Expansion frequently brought chaos in the form of much larger staff, divergent new programs, a mélange of agreements with other voluntary agencies, and recurring budget deficits. Nevertheless, what seemed chaotic to managers represented an exciting new future to nurses. Both lady managers and nurses eventually realized the inevitability of radical change in the organization of visiting nursing.[5]

Before 1900, the number of groups hiring visiting nurses grew slowly. Associations were small; many had no headquarters at all, and others used the back rooms or spare corners of dispensaries. Most were the pet charity of a small self-limited group of "lady bountifuls." In the tradition of the Ladies Benevolent Society of Charleston, South Carolina, the lady managers dominated all matters from general policy and fundraising to the pettiest decisions.[6]

Women were sought as members of the board of managers for a variety of reasons. Wealthy members were needed to support the organization, and managers with the required social connections could aid with fundraising. Male members served on financial or advisory committees whose functions were to ensure a positive relationship with the community's business, medical, and philanthropic interests. Men provided money, prestige, and advice, not time. As Annie Brainard of the Cleveland board put it, men were usually figureheads for the association, representing the highest thought and spirit of the community.[7]

The real work was done by a small group of women, those with sufficient leisure time and interest to keep the association alive. These women served on the executive committee and were responsible for supervision of the nurses. An executive committee of four or five women who remained constantly involved was given the authority to decide any questions arising between monthly board meetings. Cities were usually divided into districts, with a nurse and two managers assigned to each district. The nurse and her lady managers met weekly to review the work.[8] By all accounts, the relationship between the nurses and managers was intensely personal. The nurse clearly served as an ambassador from "one group in society to another group very removed from it."[9] This process provided a corrective influence, helping the heretofore sheltered ladies develop true wisdom and a sympathetic relationship with those living in ignorance, poverty, and disease. In exchange, the nurse was given an opportunity to review her cases with women who had seen a very different side of life—whose ideals, cleanliness, morality, and good housekeeping were exemplary. Through their personal oversight, the ladies claimed to exert an uplifting effect on the nurse's practice as she painstakingly attended to every detail.[10]

Despite such assumptions of mutual dependence between nurses and managers, little social or professional equality existed. As Katharine Tucker, Director of Philadelphia's Visiting Nurse Society, observed, most often their relationship followed the English distinction between the leisure and working classes. The workers were regarded somewhat as clerks or office boys, bound to do the bidding of the board. One lady manager suggested that a nurse could go to her manager sure of a friend. Lady managers even established savings committees, which kept back a small portion of the nurses' salary each month; nurses could withdraw funds only with the approval of the committee. While admitting that the practice was rather paternalistic, the lady managers believed the nurses were always grateful in the end. Not only did the ladies see themselves as socially superior to the nurses, they expected the nurses to defer to their judgment in professional nursing matters (Figure 2).[11]

Even in agencies large enough to require a head nurse, the managers clearly remained in charge. In 1887, for example, Philadelphia hired a head nurse, Miss Haydock, to be in charge of all those employed by the Visiting Nurse Society. The next year she was given a larger salary because of her

FIGURE 2. Lady Manager of the Boston
Instructive District Nursing Association, Early 1900s.
Instructive Visiting Nurse Association of Boston,
Boston VNA/IDNA/CHA Collection,
Department of Special Collections, Boston University.

growing responsibilities, which now included supervising the nurses, teaching students, and handling routine office business. The broad responsibilities of the head nurse rapidly eroded, however. By May 1888 several ladies had been appointed to a committee on nurses to superintend the nurses directly, to pay them monthly, and, as far as possible, to ration their work.[12] Within a month, in the absence of any input from Miss Haydock, the committee established a set of rules for the nurses dealing with personal cleanliness and behavior, hours of work, records, duties of the head nurse, and monthly meetings between nurses and managers. The nurses accepted the rules without objection. At the same meeting, the managers made clear their intention to increase their control over the nurses' practice. Miss Haydock would make decisions only about unimportant cases and would leave the disposition of more doubtful ones to the committee on nurses.[13]

Not completely satisfied with its control of the staff, the board decided that the nurses would live together under the charge of an experienced matron and under the constant supervision of the managers. The announcement caused much distress and confusion among the staff and, as expected, increased turnover. Haydock was asked to stay on under the new regulations in any position she wanted, but she declined the offer, thus terminating the first experiment with a head nurse for the Philadelphia Visiting Nurse Society.[14]

A year later (in October 1889), the ladies were clearly in charge, and Philadelphia was still without an efficient head nurse. By July 1890 an even more restrictive set of rules was enacted, perhaps in response to a specific incident or simply owing to fears about the nurses' behavior. These new rules required nurses to be neat, to treat patients with utmost gentleness, to work when on duty, and to accept no liquor from patients. House rules were provided, specifying everything from hours for meals and visitors to the time for turning out lights.[15]

A nurse's failure to comply with the rules or to provide appropriate care meant dismissal. Nurses were asked to resign for a variety of reasons ranging from an unpleasant personality to unsatisfactory performance. On one occasion a nurse complained about the Philadelphia board's decision to let her go, claiming her dismissal had been on personal grounds. The board investigated her charges and concluded that she had not neglected her patient, but it dismissed her instead for "incompetence." Displeased by the whole affair, the board then ruled "that the form of dismissal of every nurse be dictated by the board and that each nurse be engaged with the understanding that if she is dismissed as not being suited to the work, no further reason shall be asked." A lesson had been learned and the board's authority would not be undermined.[16]

Dismissal was apparently not an uncommon form of education for young visiting nurses. Mary Gardner, a member of a prominent Providence, Rhode Island, family and the first head nurse of that city's visiting nurse society, frequently chose the precarious relationship between nurses and their managers as a theme in her many writings. In *Katharine Kent,* her first novel, she described a young nurse, much as she later described herself, as lacking knowledge and experience but "eager to blaze a trail." In the novel, Katharine's enthusiasm hampered her success in her first job because, failing to recognize the value and knowledge of the board members, she acted without regard for their opinions. Wisdom was thus acquired only through much self-searching and after the board had decided to replace her with an older nurse. A family friend later comforted Katharine: "'There is nothing in your story that spells permanent disaster but there is a good deal that must be changed before you can hope to achieve success or carry your work into a better future.'"[17]

The story of Miss A. E. Beer, the first superintendent hired by the Instructive District Nursing Association of Boston, was similar. She was hired by the board in May 1900 to "see that the best work was done and to understand the relative working condition of each district."[18] Beer approached her assignment with enthusiasm and by November had visited all the nurses, reviewed their record books, and listened to their oral reports. Her summary to the board made numerous problems all too clear. She did not approve of the nurses' cotton bags, which she thought unsuitable for district work; she found the office closets were dusty and not systematically organized, the record books were not neatly kept, and the aprons were dirty. Nor was she very impressed with the staff, especially their lack of attention "to the details of district nursing and in some cases their personal appearance." Some were careless about disinfecting their hands, and some were "Sallies too old to do justice to the work." She even suggested that the society's agreement to send nurses to accompany the dispensary doctors on their morning rounds was a waste of time.[19] The board ignored Beer's impressions and recommendations, and she resigned to marry the next year. Her replacement was Martha Stark, already a member of the staff, who evidently combined the ability to supervise the staff with the wisdom to avoid criticizing longstanding traditions. In contrast to Beer's short tenure with the association, Stark's lasted eleven years.[20]

As visiting nurse associations grew, the acquisition of a staff of twelve nurses, a business office, and an occasional nurses' home seemed to mark the point at which most lady managers were unwilling or unable to sustain control over the organization. The usual solution was to hire a head nurse or matron to supervise the daily work of the staff. Even so, the ladies quickly found it difficult to relinquish control of any aspect of the work and frequently became overbearing in their struggles to find an appropriate balance of managerial control. On the other hand, inexperienced head nurses, through their excessive enthusiasm and criticism, often gave the appearance of disregard far beyond work issues and antagonized the board. Compromise was uncommon; either the head nurse conformed to the will of the board or she was replaced by someone more cooperative.

A Decade of Change

Despite these struggles, most visiting nurse associations could not continue as the sole domain of the lady managers. With more staff and new programs, internal and external relationships began to go awry. Record systems were inadequate, supervision superficial, and programs overspecialized; no matter how reluctantly, the lady managers increasingly recognized that their beloved creations were inefficient and required radical reorganization.[21]

Most boards sought professional help. Management of their organization became the responsibility of a new kind of superintendent, a woman "of unusual training, splendid endorsement, and courage."[22] In fact, these superintendents differed significantly from their predecessors. Well educated, often college graduates, they came from good families and could be trusted by the ladies, as one Boston manager suggested, to take "hold of their problems . . . with a sympathetic understanding of the past and a clear vision of the possibilities for the future."[23]

Despite the perceived necessity for these changes, lady managers feared that lack of daily involvement would produce in nurses a growing sense of estrangement. Might the staff, queried an older Cleveland manager, come to see the ladies as "well-fed, well-housed, and well-warmed" managers who "hold the purse strings and count out the dole?" Just as surely, she believed the lady managers were doomed to experience a kind of spiritual anemia, as well as indignation at being "relegated to a role where one's money is thought more useful and desirable than oneself."[24]

Nevertheless, city by city, a pattern of reorganization and redefinition of authority was repeated. In Boston it was accomplished by Mary Beard; in

Providence by Mary Gardner; in Chicago by Edna Foley; in Dayton and then in Washington, D.C., by Elizabeth Fox; in Baltimore and Los Angeles by Mary Lent; and in Philadelphia by Katharine Tucker (Figures 3 to 6). That the outcomes were fairly consistent is not surprising, since this "ring of women," as Mary Gardner called them, regularly exchanged ideas and experiences. This same network of nurses with their lady managers would, in 1912, form the National Organization for Public Health Nursing to promote similar developments nationwide.[25] The Boston Instructive District Nursing Association (IDNA) was one of the first to reorganize, and I consider this in detail because it typifies the changes that occurred in many visiting nurse associations immediately before World War I.

The first inkling that the lady managers of the Boston association shared any concern about their methods occurred in February 1908, when Elizabeth Cordner, a member of the board, sought the advice of Dr. Hugh Cabot, a Boston physician. Cabot began his five-page response by reminding the ladies that any association dependent on private contributions must be in a position to ensure that no one could question that its work was of the highest attainable standard. This, he believed, was achieved through careful, constant supervision, and he concluded that, unfortunately, "in some respects the supervision of the nurses of your Association was ineffective." Most pointedly, he reminded the ladies that meeting with the nurses to discuss cases known only by a name failed to constitute the searching, semi-medical inquiry essential to good work. He realized a superintendent was in charge of the nurses, but he found no evidence of strict supervision over the past years. As for Superintendent Stark, he concluded, "I should not expect her to do such work as well as other parts of the work which she finds more congenial."[26]

Cabot suggested that for the most efficient supervision the ladies should hire a superintendent of considerable experience and ability in the field, who would give her whole time to overseeing the nurses' work. Because this would involve an expenditure of money that the board might hesitate to make, he also suggested obtaining the services of several young medical men "whose business it would be to make rounds with the nurses at frequent, though unknown, intervals and report directly to the Board of Managers." These gentlemen would undoubtedly have weight with the community and, thought Cabot, would make fundraising easier, provided they found the work was being done with the greatest possible efficiency.[27]

Not sharing Cabot's assessment of their methods, the ladies found other

FIGURE 3. Mary Beard. *Instructive Visiting Nurse Association of Boston, Boston VNA/IDNA/CHA Collection, Department of Special Collections, Boston University.*

FIGURE 4. Mary Gardner. *Providence District Nurse Association, Visiting Nurse Association of Rhode Island.*

FIGURE 5. Edna Foley. *Visiting Nurse Association of Chicago, Chicago Historical Society (photo: Kaufmann & Fabry Co.).*

FIGURE 6. Katharine Tucker. *Visiting Nurse Society of Philadelphia, Center for the Study of the History of Nursing, School of Nursing, University of Pennsylvania.*

matters more important. During the next three years, their preoccupations were a growing staff (from fifteen to forty-three people) and a corresponding increase in their budget (from $27,000 to $49,000). The association negotiated new business arrangements with the Metropolitan Life Insurance Company, expanded the maternity program to include prenatal care, initiated programs in contagion, day nurseries, and industrial nursing, and expanded the new training school for district nurses.[28]

But by 1911, the ladies were forced to examine the outcome of these new programs. Visiting nursing in Boston was no longer confined to the specialized programs of the IDNA; both the Baby Hygiene Association and the Health Department had their own staff of nurses. In fact, the potential now existed for Boston families to be visited by up to ten different "types" of visiting nurses. The IDNA managers concluded that "there are today too many unrelated groups of nurses in Boston. Among them there is much duplication, waste of time and money, and the public, which we are trying to help, is in reality the sufferer."[29]

Michael Davis, Director of the Boston Dispensary, shared the managers' concern that the nurses' work was too specialized. He believed specialization produced the highest efficiency only when workers "are under a strong, broad and intelligent supervision and when the specialists have a good background of training behind their particular skill." Otherwise, he reminded the ladies, the result was a number of specialists working more or less independently or without the proper correlation. Not only was this a waste of time, it would also fail to effect the desired results. According to Davis, what the association needed first was a more authoritative, centralized organization and, at least for a time, less specialization. Eventually the managers should hire "a higher type personality or at least provide more training of the rank and file worker."[30]

Ultimately, the IDNA decided to reorganize and began to search for the right nurse to implement its plan, which incorporated ideas of professional supervision, centralization, and training.[31] The lady managers believed they had the solution to something more than just a local problem; they were convinced it would serve as a national example. But the search for a new superintendent immediately encountered difficulty and before it was over, one year later, the Boston managers had been turned down by some of the most innovative leaders in the field, including Ida Cannon, Edna Foley, Ella Crandall, and Ellen LaMotte.[32]

That the board would choose to vigorously pursue someone as radical as

Ellen LaMotte suggests either its desire to create a totally new order or, perhaps, its desperation. At the time, LaMotte was the only female department head in the Baltimore Health Department. Although her letters of reference consistently praised her unsurpassed ability to handle a difficult situation, her quickness of mind, her power of literary expression, and her original ideas, they all ended on a cautionary note suggesting, in essence, that the same qualities that made LaMotte one of the ablest women in the nursing profession might make her not totally acceptable to the Boston IDNA board. Their concern included her very pronounced character and strong opinions, her tendency to become bored, careless, and indifferent if the work became routine, and the possibility that she might want to do everything the Johns Hopkins way. Finally, she was known to be "a rabid suffragist."[33] Apparently willing to take its chances, the board offered LaMotte the job. LaMotte, however, was unwilling to leave her Baltimore position, feeling it was still too unsettled.

Finally, in November 1911, Ella Crandall recommended Mary Beard to the Boston managers, but they hesitated to hire her because of her "unpreparedness" for undertaking such "a big social and nursing cause." No one "better fitted" was at hand, however. Crandall, anxious to see the position filled and thus save herself any "repeated solicitations" from the managers, assured the board that Mary Beard was the right person for Boston. Regardless of her motivation, Crandall did claim to be very impressed with Beard, "her appearance, voice, manner, mental attitude, and mental grasp" and especially her being a "New England lady born and bred."[34]

Beard did, in fact, have several years of experience as a head nurse with the Waterbury, Connecticut, Visiting Nursing Association. Its president, the Reverend John Lewis, thought that on the whole, Beard would be well suited for the Boston position because of her ability, self-confidence, and personal charm. He noted that she did much better when in command than as a subordinate, managed other people's work better than her own, and was not a good financier or good at keeping accounts. He assured the board that she would do admirably if she had a secretary, adding unequivocally that "she needs one."[35]

When Mary Beard began her new position in February 1912, she was confronted with twenty-five years of tradition, twenty-seven lady managers whose expectations included the elevation of their association to "its proper place in the community and the country," and forty-nine nurses caught in the middle. By April, she initiated a plan of reorganization so complete it produced an institutional revolution. Although most of the nurses remained on the staff, some found it difficult, as Beard put it, to "fall in loyally with the changes."[36]

Neighborhood Nursing

Mary Beard's plan called for consolidating the work of many nurses so that each would no longer specialize in one type of problem but would become a general practitioner—what Beard called a neighborhood or community nurse. Consistent with this neighborhood plan, seven branch stations were created, each with a supervisor for the "executive work" (advising and supervision) and a nursing staff to meet neighborhood health needs.[37]

Beard envisioned the nurse as a community consultant for families in which "illness has produced poor living conditions or those in which poor living conditions have produced illness." Eventually, she predicted, "all the people of a neighborhood, regardless of income, would employ our nurses and each patient so cared for will feel a personal responsibility to raise the money necessary . . . in order to supply all the nursing the community requires." If successful, Beard's plan would shift the focus of visiting nursing from care of the sick poor to a service for the whole community.[38]

Recognizing that much "educational work" was required before her plan could succeed, Beard devised an attention-getting strategy. The nurses' uniforms were changed to symbolize (and advertise) that they were now the neighborhood's nurses. The principle of collecting a fee, however small, was emphasized, and the association's "scale of prices," ranging from nothing to 50 cents a visit, was conscientiously publicized. Furthermore, a more "scientific" and accurate method of keeping records was established, making it possible to better document the good work being done and the relationship of that work to national health problems.[39]

During her second year with the association, Mary Beard's major concern was upgrading staff by replacing all the temporary nurses with twenty-two new nurses who had completed a postgraduate course in public health nursing. These replacement nurses, she assured the board in 1913, "want to do this sort of nursing more than any other work, are full of enthusiasm and have made sacrifices to obtain a special education for it." In 1914 she reported that the nurses were responsible, independent authorities and that the amount of work accomplished had increased. All that remained was the updating of the "Rules for the Nurses" and creation of written standards describing proper techniques for each type of nursing care. These standards would be used to grade nurses, thus ensuring elimination of waste and efficient fulfillment of the association's ends.[40]

When Beard's transformation of the IDNA was complete, it too would be graded. According to Beard, one had to be willing to test one's efforts and results with courage and a sincere desire to learn the truth, either positive or negative. Such a test should, first, make comparisons; second, determine whether any historic developments had been achieved; and finally, analyze and study the work in relation to past achievements and future needs.[41]

The formation of the National Organization for Public Health Nursing made the "test by comparison," as Beard put it, "both delightful and easy," for at the annual meetings members of the IDNA's staff and board could talk with representatives from visiting nurse associations from across the United States and Canada. Mary Beard, Mrs. Codman (a board member), and four staff nurses attended the first meeting, held in Atlantic City in June 1913. Not surprisingly, they found that many shared their concerns about standards, records, statistics, and the need to extend their work to "all people."[42]

Beard reported in the annual report for 1913 that, as she remembered it, the central idea expressed at these meetings was that of neighborhood nursing. In fact, Boston's official decision to abandon specialization in favor of a generalized service placed its visiting nurse association in the forefront of a debate that remained unresolved for years.[43] On the whole, however, the comparison proved favorable. The Boston association had clearly attained a position of prominence.

Mary Beard saw neighborhood nursing as an important contribution to the health of the community and chose to interpret fees collected as an appropriate index of success. Consistent with this criterion, she included a "fee study" in her annual reports, but left it to the reader to decide whether these results indicated success or failure. Based on Beard's data and criteria, most readers would probably have concluded that neighborhood nursing was not a great success—since the percentage of patients paying fees remained stable at about 24 percent. Data available in the treasurer's reports suggest a somewhat more positive picture, as the percentage of total budget received from fees increased from 3 percent ($1,521) in the year preceding reorganization to 9 percent ($6,333) by 1915.[44]

In terms of productivity, neighborhood nursing was clearly a success. Between 1911 and 1915 the number of yearly visits increased by 15 percent (16,922) and the number of patients receiving care increased by 41 percent

(4,100), while only three nurses were added to the staff. For each nurse, this increase meant caring for ninety-eight more patients a year. In a financial sense, the outcome was not entirely positive because the nurses, who were better educated and working harder, were also better paid. But the additional $17,000 in wages seemed to cause the managers little distress. The board members consistently encouraged Beard's plans for expansion, reassuring her from the onset that they planned to think in terms of work rather than money. Although they assured her the money would be there to do the work, they incurred, and eventually covered, a deficit of about $3,000 each year.[45]

At the end of four years, Mary Beard, with her board's support, had created a profoundly different organization. Its approach to finances had shifted from total dependence on donations toward the patient as a source of income. Gone were the lady managers' naive business practices, replaced by the more professional approach of nurses committed to eliminating waste, maintaining reliable records, and consistently producing the highest quality of nursing care. The most dramatic change, however, was the altered relationship between managers and nurses. As the nurses began to dominate, the managers retired into the background to such an extent that the supervisors began to complain they no longer knew their names.[46]

These changes in Boston were only the first of many such transformations across the United States. Not surprisingly, home care's newfound interest in efficiency and scientific management was not an isolated occurrence. Across the country, experiments in the management of businesses, hospitals, and even household tasks promised higher standards and better outcomes.[47]

Part II

The Work and Reality

CHAPTER 3

"Treatment of Families in Which There Is Sickness"

O n the surface, tenement slums, immigrants, and the "dangerously ill" appeared to create an unstable and uncontrollable situation in poor neighborhoods at the turn of the century. To doctors and nurses who practiced door-to-door, however, these neighborhoods were simply a "large outdoor hospital." The types of physicians who cared for the poor varied. Some worked with dispensaries, others were lodge doctors who contracted with one or more mutual benefit societies serving immigrant communities, and others were simply private physicians. Regardless of whether the sick poor distrusted physicians or had great faith in their skills, poverty meant seeking medical care only for dire emergencies.[1] Likewise, the poor relied on family members for care as long as possible before seeking the charitable assistance of visiting nurses, occasionally paying a little for a nurse's caregiving. For the nurses, the neighborhood was the ward where they directed and supervised patient care. As in the hospital, a system of bedside notes and case records existed, as well as an established etiquette among physician, nurse, and patient. Unlike in the hospital, in the patient's home the family was in charge. Doctors and nurses were visitors who quickly came and went. Depending on their expertise and interpersonal skills, doctors and nurses were trusted, welcomed, endured, or seen as intruders. Family's caregiving skills were often a reflection of this triangular relationship between nurse, doctor, and patient.[2]

Initially, doctor-nurse relationships were dictated by the lady managers. The Boston visiting nurses of the Instructive District Nursing Association (IDNA), for example, worked under the physicians of the Boston Dispensary. The nurses were the "fortunate medium between the doctors and the poor . . . giving grace to all the work."[3] Each nurse was assigned to a section of the city, and each district had a physician with whom the nurse worked. At some regularly appointed place and hour, the nurse met the physician to receive new orders, report on the previous twenty-four hours, make home visits with him, and attend operations when necessary. Visits to new patients or very sick patients were usually made with the dispensary physician at the beginning of each day, after which the nurse continued her rounds alone. Initially, less than half of the nurses' visits were made with physicians. The nurses also relied on local apothecaries to hold messages, orders left by doctors, and families' requests for help.[4]

Over the years, gradual changes occurred in the routines of nurses and dispensary doctors. As long as harmony was maintained, the lady managers rarely intervened, but the occasional deviation provoked them into action. For example, a poor distribution of cases and "problems of personality" resulted in communications with the Boston Dispensary's superintendent; a district physician's attempt to alter the morning routines led the lady managers to send a copy of *The Nurses' Rules for Work* with the appropriate sections underscored. When doctors' irregular schedules kept nurses waiting, the ladies decided the nurses could wait for no more than twenty minutes. Physicians' attempts to alter visiting loads to suit their convenience were considered "high handed." Routinely, such problems were handled by the managers, not the nurses.[5] Some physicians, however, were reported by the nurses and removed by the dispensary for having little compunction in the care of patients, tendencies to favor patients of certain nationalities, and a preference for those who paid a little for their care.[6]

For their part, most Boston Dispensary physicians accepted the assistance of the visiting nurses, although one reportedly decided to hire his own nurse. While the physicians agreed to work with the nurses, they confided to the lady managers that "the constant danger with trained nurses is that they shall usurp the doctors' position and prescribe for patients." The physicians described some nurses as a great success—both professionally and personally. Others were said to be good nurses but talkative, or to have exceptionally bad feet or bad underarm odor. By the turn of the century, however, the dispensary doctors were reporting complete satisfaction with the nurses' work. Once the doctor had "satisfactorily" diagnosed a patient's illness and given "directions in regard to treatment," the patient was left in the nurse's hands. The physician returned only when the nurse needed him.[7]

Within a few years, private physicians also began referring patients to the IDNA. The ladies decided they would accept referrals from doctors included on the Massachusetts Medical Society and Homeopathic Doctors List. Like the dispensary doctors, these physicians also received mixed reviews. Often referred to as "local doctors," some were quite acceptable and others irascible. Referral relationships also developed quickly with several Boston hospitals for follow-up visits to patients recently discharged from the hospital and those referred by the hospitals' outpatient departments. Eventually, families would call for the nurses themselves.[8]

Variations on these themes developed around the country. Initially, the lady managers established and orchestrated relationships with the medical community because the need for home nursing was not always obvious to physicians. For example, in Philadelphia the ladies reported that until they enlisted the help of some leading physicians, no one seemed to need the Visiting Nurse Society's services. While the society reported that Italian doctors "are not altogether friendly to the visiting nurse, fearing that with a nurse in attendance they will not be needed so frequently,"[9] the proportion of Italians seeking the care of a visiting nurse nevertheless increased dramatically, accounting for 25 percent of the cases by 1904. Trust would eventually come, the nurses found, but often had to be cultivated.[10]

Connecting with physicians of all types was greatly simplified by the growing availability of the telephone. By 1906 in Boston, for example, phone orders made the complicated requirement for daily rounds obsolete.[11] The opening of neighborhood offices further extended the influence of the visiting nurse, while reducing her dependence on any particular physician or institutional relationships. By 1915, Lillian Wald, the founder of public health nursing, could write that one of the fundamental principles of nursing the sick at home was that access to nursing care should not depend on patients' connections with certain physicians or institutions. Nurses, according to Wald, should respond to calls from individual patients and physicians unencumbered by red tape or formality.[12] Now, families and nurses were referring as many patients for care as were physicians. By the 1920s, nurses were working with hundreds of different doctors.[13]

As nurses took over the leadership of visiting nurse associations (VNAs), they also took control of negotiating relationships with the medical community. In nurse-author Mary Gardner's fictionalized reminiscence of this recasting of roles, Dr. Waldron reminds his young visiting nurse colleague, Katharine Kent, "'Your profession and mine are passing through a period of transition . . . in relationship one to the other.'" Thanks to nurses' "good professional education," they were moving from handmaiden of the physician to co-worker. This new status of cooperation was confusing to some physicians, he admitted. Visiting nurses were reformers, and reformers were "the very devil for non-reformers to work with." Dr. Waldron hesitated. " 'Then too, you know, Miss Kitty, there are nurses and nurses.'" From the author's viewpoint, "kindliness, understanding and a selfless attitude on both sides" would create "a united and powerful good in the world."[14]

As might be expected, nurses who increasingly found themselves alone in patients' homes without direct medical oversight simply took charge of what was happening. The telephone allowed them to function even more independently. By calling in to the office in the morning and at noon, they could set their own schedule free of medical oversight. They responded to all new calls from any source, including families, but were expected to continue care only "in conjunction with the physician in charge." Nurses relied on "standing orders" until a physician could be secured or when the attending physician had not left any orders. Standing orders covered frequently encountered situations such as new patients and patients with elevated temperature, infantile diarrhea, burns, ear discharges, infantile convulsions, chronic ulcers, postpartum hemorrhage, dressings to be changed, and emergencies requiring first aid. Nurses' Rules often stated that nurses were not to refer patients to hospitals, clinics, or specialists without the consent of the family physician, but physicians regularly accused them of this "evil" practice.[15]

Standing orders and nursing procedures might be written by a medical committee appointed by the local medical society, a group of "interested" local practitioners, or physician representatives on the VNA's governing board. The point was that in order to practice, the nurses needed the approval but not the presence of local organized medical groups. To avoid trouble and criticism, the chosen medical advisors reflected local medical opinion and practice patterns. Most VNAs eventually established medical committees or appointed respected physicians to arbitrate the inevitable disputes between nurses and doctors. Local physicians were always to be consulted before outsiders, that is, those "outside the district," especially specialists.[16]

Under less physician control than hospital nurses, visiting nurses exercised a different level of authority—what historian Susan Reverby has called "care with autonomy." Although these nurses carried out physicians' orders, their working conditions allowed them to discriminate among doctors, co-operating with those who were working for a higher standard of public welfare, while standing in only "remote and casual relation with those who have no such aims or desires."[17] Even more intimidating to some physicians was the nurses' easy access to the homes of potential patients. While never "diagnosing," nurses did believe it was their business to select patients requiring diagnosis and to refer them appropriately. Although the actual selection of a physician must be left to the family and the nurse simply shared an approved list of physicians practicing in the neighborhood, it is unlikely such "sharing" occurred in the absence of the nurse's opinion.[18] According to historian Alan Kraut, families in ethnic communities commonly hired and fired numerous physicians, even in times of crisis.[19]

Ellen LaMotte, an early leader in the field, noted the singular independence of nurses who cared for tuberculosis patients. No special orders were written; the doctor knew what should be done and the nurse knew how to do it. Extensive discussions were, claimed LaMotte, unnecessary. Patients could go for months without seeing the doctor or could even change physicians, without significant impact on their care. The nurse was in charge of the patient's care, remained through the long months of illness, and moved with the patient from doctor to doctor. Some physicians reacted to this behavior with antagonism and opposition, but according to LaMotte they were simply "holders from a passing regime." In such cases the nurse simply proceeded with her duty, even if at cross purposes with the physician.[20]

Home Nursing

Most visiting nurses worked six days a week, making only emergency visits on Sunday. A nurse made between eight and twelve visits on a normal day, staying in each home for thirty minutes to two hours depending on the severity of the problem. Most patients received an average of eight to ten visits before discharge. A nurse's typical day would begin at 8:00 a.m. with visits to the patients considered seriously ill. Her day ended eight to ten hours later with a return visit to these same patients. The rest of the day was devoted to care of patients with problems such as varicose ulcers, bronchitis, fevers of various origins, abscesses, and assorted chronic and infectious diseases. Good nursing care included ensuring that room air was fresh and wholesome; keeping the patient, patient's bed, and sick room clean and the surroundings quiet; establishing a routine for meals and medicines; making skilled observations of the patient's vital signs and condition—carefully recorded or communicated to the physician; and taking appropriate measures to prevent the spread of disease.[21]

Nurses carried the "tools of their trade" in a black bag. The contents usually included instrument case, scissors, forceps, probe, thermometer, soap, bowl, nailbrush, towel, apron, bandages, brandy, and tape.[22] Advancements in medical technology such as portable sterilizers and oxygen containers were also introduced into patients' homes by the visiting nurse. As one medical authority suggested, the visiting nurse served as the "relay station, to carry the power from the control stations of science, the hospital, and the university to the individual homes of the community" (Figure 7).[23]

Much of the nurses' efforts focused on alleviating patients' distressing symptoms through the use of an armamentarium including glycerine for parched lips, cold compresses or icebags for headache, and weak mustard plaster or a little bruised mint in brandy over the stomach for vomiting. Patients with fever were treated with a "temperature bath," and a cool atmosphere was created by pouring cold water on the floor or by hanging sheets wrung out in ice water around the room. Patients with bronchitis or pneumonia received similar care, and the nurse also rubbed the patient's chest with camphorated oil and dressed her or him in a cotton jacket or covered the chest with hot flannel. Rheumatic joints were bandaged, children with measles were bathed in saleratus water, and the danger of contagious disease was abated by hanging up sheets dipped in a carbolic solution.[24] The nurses taught these and certain basic emergency interventions either to the family or to a neighbor to be carried out in the nurse's absence.

In addition to providing this "general care," visiting nurses often found it necessary to contribute clothing, nightgowns, wheelchairs, urinals, carts, crutches, and food essential to their patients' recovery. They also provided volunteer babysitters so that mothers could take sick children to the hospital or clinic for care. When necessary, nurses provided visiting housekeepers to do domestic chores for families considered "unable or unreliable." Hiring housekeepers to clean houses saved the nurse's strength for actual nursing care, for which she was really employed.[25]

Occasionally, visiting nurses were even called upon to treat the pets in patients' homes. One canary with asthma was cured by a change in climate (removal from a smoky kitchen); in another instance the nurse claimed to have successfully treated a monkey with measles. For the visiting nurse, healing the sick and spreading the gospel of cleanliness could not be narrowly defined.

FIGURE 7. Visiting Nurse Calls on a South Philadelphia Family, 1886.
*Visiting Nurse Society of Philadelphia, Center for the Study of the History of
Nursing, School of Nursing, University of Pennsylvania.*

In some cases, such complex undertakings were seen as not only impossible but inappropriate.[26] To survive, suggested one nurse, you must conserve your own "force and not recklessly scatter [your] time, strength, and service." The wise nurse followed the dictum of "a few patients well treated [is] more satisfactory than many just cursorily looked after."[27] Thus, survival required "a stratum of strong good sense" that enabled the nurse to accurately assess which cases to follow and which to avoid. Climbing five flights of stairs with a full day of serious cases awaiting her, only to find the patient away or "incorrigible," was usually the only lesson a nurse required to understand which patients to bequeath to the sole care of the doctor or the family. Representing anywhere from 9 to 30 percent of all cases, such patients were described in annual reports as terminated to "home care" after a single visit—that is, the family was able to provide care without the nurse's assistance. From a contemporary perspective, it is interesting to note that the earliest use of the term *home care* by visiting nurses was in reference to family caregiving.[28]

"Who Shall Nurse the Sick?"

At the opposite end of the illness continuum were patients too sick to be tended adequately in one or two visits a day and then left in the family's care.[29] At the end of the nineteenth century, these typically included patients with typhoid fever, diphtheria, scarlet fever, or other serious diseases, who required constant nursing care but for whom hospitalization was inadvisable or not a possibility. Special or emergency nurses were provided to stay with patients for up to twenty-four hours after major surgery and could be sent to do night work when necessary. VNAs would pay for continuous nursing (when the patient's condition was not chronic) for up to three days if so ordered by the doctor. The need for these nurses varied dramatically, from twenty-five to five hundred times per year.[30]

Acknowledging one of the great difficulties caused by caring for the sick at home, the Chicago VNA used emergency nurses for cases in which the "bread winner of a family is being kept at home to care for the patient, in all cases where the life of the patient depends upon the constant care of a nurse, and in cases where relatives are worn out with care of the patient."[31] In some instances, emergency nurses were actually non-nurse attendants supervised by the fully trained graduate nurse in charge of the case. Usually, women hired as attendants had some experience in caring for the sick, were certified by the association, and, most important, were willing to stay in "the poorest accommodations possible and endure great privations." Paid $1 per day and $7 to $12 per week, attendants were much less expensive than the trained nurse. This practice of "substituted care" was reluctantly discontinued after considerable criticism from physicians and "others" for the use of untrained personnel. On the other hand, use of VNA staff on a rotating basis to provide emergency care was never popular. Nurses were unwilling to stay in the homes of poor patients; furthermore, night duty made it nearly impossible to maintain normal working conditions. With no other options, VNAs referred cases requiring more extensive care to local nurse registries. Families were expected to pay as much as they could for this service. Some VNAs had special funds to help families unable to pay. After 1920, this so-called emergency care was rarely mentioned and seldom required. Home care had become a service provided between 8:30 a.m. and 5:00 p.m. and was no longer appropriate for patients sick enough to require continuous care or night calls. These patients were increasingly cared for in the hospital.[32]

The nursing of chronically ill patients raised similar questions about what care was appropriate and who should provide such care. By the early 1890s, physicians of the Boston Dispensary were using attendants rather than nurses to provide care for the chronically ill. The IDNA managers lacked enthusiasm for this idea, but it apparently pleased the physicians and so the ladies of the IDNA kept their concerns to themselves.[33] The matter of who should nurse the chronically ill reappeared in the first decade of the new century, when proposals to transfer the care of the chronically ill to less expensive attendants again challenged nursing's unresolved agenda of professional self-definition. Still struggling to convince the public that expert nursing care was important, nurses felt called upon to resist the potential threat posed by care of the chronically ill by non-nurse attendants. Although nursing cloaked its opposition to attendants in the rhetoric of guarding the public through high standards of care, all recognized that many chronically ill patients could get along with less sophisticated care. At a New York Academy of Medicine meeting, Annie Goodrich articulated the position of nursing's leadership: "If there is another body or class of worker needed, it will come into existence; we believe, indeed, that such a class is here and is only waiting to come into an orderly existence for the field of the more important worker, the nurse, to be developed . . . If the doctor and the family are satisfied to relegate their sick into her hands, well and good. Our responsibility ceases. Our point has been made when the line of demarcation is clear."[34]

The challenges posed by the possibility of less skilled workers assigned to care for the chronically ill provided the impetus for visiting nurses and their lady supporters to organize themselves. By June 1912, they had created the National Organization for Public Health Nursing.[35] Of course, inventing an affordable and safe approach to caring for the sick at home still required resolving the problems of which patients should be cared for at home and the level of training needed to provide that care.

Although it was understood that many of the visiting nurses' patients were ill enough to warrant hospitalization, the nurse's special role was to ensure that her patients never became that ill. The nurse's responsibilities also included identifying patients who "properly" belonged in the hospital rather than at home—and in her caseload. The patient's condition, duration of illness, and financial situation and the condition of the home surroundings were used to determine the appropriate locus of care. Suitable cases for the hospital included patients with "brain troubles" demanding quiet surroundings, contagious diseases endangering the family and other tenement dwellers, or fractures and injuries requiring apparatus not available at home, and patients requiring "climatic care," continuous care, or operations under aseptic conditions. Despite this ever-expanding list for hospital care, on average only 6 to 10 percent of patients in the nurses' caseloads were admitted to the hospital.[36]

Overcoming the "hospital prejudices" of the poor often proved a challenge. Visiting nurses were careful to establish themselves as good neighbors whose judgment could be trusted. When their reputation occasionally failed to produce the desired results, they used more extreme measures. Lillian Wald's most extreme form of persuasion was leaving the patient alone "for a short time that she might feel how dreadful it was." When successful, nurses reported accompanying their patients to the hospital, visiting the hospital to provide oversight of their care, and visiting the family for the duration of the hospitalization.[37]

After careful removal of "comparatively worthless cases" and those better cared for in the hospital, what remained was, as one visiting nurse suggested, "our legitimate work."[38] Like Wald, some claimed that any patient in bed needed a visiting nurse, but others were more cautious. Philadelphia's Visiting Nurse Service, for example, claimed sudden acute illness, simple injuries, "diseases of dirt" such as bedsores, and maternity cases as within its appropriate realm of responsibility. Patients with chronic and incurable conditions such as cancer or rheumatism, which were seldom treated in hospitals, were their responsibility as well. During seasons of "comparative neighborhood healthfulness," the nurse's domain extended to follow-up of minor illness among children, teaching families about simple nutritious cooking, and keeping children clean and free of vermin. Despite these seemingly compatible visions of the nurse's work, few could agree on the relative value of hands-on nursing care as opposed to educating patients and families in the ways of health. By 1904, the proper balance between these two activities had already become, as it would remain for years, an unresolvable issue.[39]

Statistics reported in annual reports provide interesting details about patients and their problems. Boston's IDNA kept complete data during this early period, and its experience is probably typical of most larger visiting nurse associations. In Boston prior to 1910, the vast majority of patients were women and children. While most patients (70%) were treated for acute medical problems, 14 percent had chronic illnesses, 11 percent were surgical cases, and only 2 percent were obstetrical cases. In addition to treating pa-

tients, nurses continued to believe in the treatment of the patient's surroundings—typically in the form of referrals to health departments for fu-
migation. Continuing a tradition begun by the ladies of Charleston, VNAs
regularly reported patients' status at discharge in their annual reports. Most
patients were listed as well or improved, a small percentage (5%) continued
to need care, and fewer than 5 percent had died before discharge.[40]

By 1909, patients seen by most visiting nurses were categorized as surgical, medical, obstetrical, contagious, or tubercular cases, although not every
agency accepted all such cases. The result was a muddled, unpredictable, and
varied assortment of services depending heavily on the sponsoring organization and prevailing local practice patterns. Most agencies (90%) cared for
tuberculosis patients, but many agencies also visited patients categorized as
medical (70%), surgical (69%), and obstetrical (62%) cases. Despite the
prevalence and importance of contagious diseases, fewer than 15 percent of
agencies cared for patients with these conditions.[41]

Nurses and the "Dangerously Ill" and Their Germs

Consistent with medical opinion, tuberculosis was not categorized as a contagious disease. Although the disease was communicable, the danger of
nurses spreading tuberculosis from house to house was considered minimal.
In 1909, the diseases viewed as contagious were primarily measles, scarlet
fever, diphtheria, whooping cough, typhoid, and smallpox.[42] Tuberculosis
was also considered a "diathetic" disease, one with both physiological and
environmental causes, a "house disease" of the very poor. Because tuberculosis caused about 10 percent of all deaths—as much as 15 percent in some
cities—its eradication required the cooperation of thousands of patients
and families. Even with four hundred tuberculosis hospitals and sanatori-
ums across the country, the campaign against tuberculosis occurred mostly
in the community. Not surprisingly, 90 percent of visiting nurse organizations in 1909 cared for tuberculosis patients.[43]

Visiting nurses did not underestimate the danger and prevalence of tuberculosis, and the role of nurses in protecting the public's health was regularly celebrated. In a frequently repeated story used as visiting nurse "propaganda," a nurse visits a family of six, with a seventeen-year-old daughter
who helps support the family through her small candy store. On this visit
the nurse discovers the presence of tuberculosis in the family. The nurse describes to her listeners how the odors of "the making of candy, the sleeping,

eating, washing of the family" and the odors of the street mingle in the same basement store with the patient's consumption. The story's customary conclusion was, "Think of all the tuberculosis bacilli presented with each package of candy." The message—that the diseases of workers who processed food or made clothing in their filthy tenement homes could spread to decent, clean, and respectable citizens—was clearly intended to serve as a powerful incentive for renewed support of the visiting nurse's mission to eliminate uncontrolled illness among the poor (Figure 8).[44]

Paradoxically, although visiting nurses led the war against tuberculosis, they were conspicuously absent in the homes of patients with contagious diseases. It was generally agreed that such patients should be cared for in municipal hospitals designated for that purpose, but sufficient numbers of beds were rarely available and most patients were unwilling to enter these facilities. Reporting and isolation requirements were traditionally enforced for those living in boarding houses, hotels, and tenements, but rarely for the wealthy. If a case was reported, an inspector from the board of health left printed instructions and "some orders in regard to isolation." No follow-up visits were made, and those quarantined at home were left to the care of family. Isolation and disinfection procedures in tenement buildings were financially devastating to families dependent on home manufacturing, since the work was stopped and goods could not be sold until cure and disinfection were accomplished.[45] In many cases, the father of quarantined children was forced to live outside the home so that he could continue working. All well children in the family were excluded from school for a period of ten days to several weeks. In crowded tenements, the mixing of sick and well often resulted in what one visiting nurse described as a vast and hopeless jumble.[46]

Starting in the 1890s, Boston lady managers were concerned about the "unusual persistence" of contagious diseases. Believing that legislative control was insufficient, they formed a special committee to study the problem. Nurses were instructed to collect data on what might be done to lessen the danger of contagion.[47] Still troubled, the lady managers in 1909 examined the effectiveness of isolation "among careless and ignorant people." Confirming their fears, this study found that in about one-quarter of reported cases of contagion, patients were not properly isolated. Nevertheless, the health department remained unwilling to hire nurses to make follow-up visits to these patients.[48]

In her popular book *The Public Health Nurse in Action,* nurse-author Marguerite Wales described a typical all-out nursing assault on contagious

MURDER!!!

FIGURE 8. Illustration from Chicago Department of Health Educational Series, 1918. *City of Chicago, Department of Public Health.*

disease. A story titled "Caught in Time" tells of the nurse sent to a neighborhood school to "inspect" Alice, an eight-year-old child with a sore throat. Fearing something catching (diphtheria), the nurse sends Alice home and reports her case to the health department as suspicious. A visit from the health department's physician confirms the diagnosis, and Alice is isolated and reported as a carrier. The nurse's visit provides reassurance to Alice's mother, demonstrates necessary care, but, most important, focuses on how to prevent the spread of "this scourge." Mother and nurse then reconstruct Alice's activities over the past five days, hoping to identify the source of infection. The health department's "extensive investigation" moves on to Alice's school, where the nurse conducts a schoolroom inspection and cultures the throats of several children. While no other cases of diphtheria are identified, the nurse's "prompt action," recognition of symptoms, and establishment of routine procedures are credited with protecting the children in Alice's school. Nurse Wales notes that the policy of calling the patient's private physician was in most cases considered the best practice. But in "some situations"—such as this one—the nurse was permitted to first notify the health department, place the patient in isolation, obtain a diagnosis, and only then call the private physician. Apparently, these judgment calls reflected both the degree of danger to the public and the nurse's attitude toward the family. Historian Alan Kraut contends that poor families often saw the health officers and visiting nurses who had the power to quarantine as unwelcome intruders, feared and resented. This was no doubt true in some but not all instances.[49]

Given the anxiety about cross-infection, agencies providing direct care to patients with contagious diseases often maintained a separate staff for these cases. The cost of this expensive undertaking was explained by the need to pay these nurses a higher salary and often to send them greater distances, and for the nurses to spend more time than usual in making visits. Safety precautions ensuring that nurses did not spread disease were essential, no matter how expensive.[50] The approaches to contagion by some agencies were often rigorous and time consuming. In Chicago, for example, the following procedure was followed for patients with contagious disease requiring personal care: "If, however, the personal care of the visiting nurse is needed, she must visit all her other patients first. She removes her out-door garments before entering the sick-room, putting on, if possible, her rubber coat, leaving outside also her bag of supplies. After giving the patient all necessary care, she returns to her room, sprays her coat and bonnet with bichloride spray, washes her hair with boracic solution, takes full bath and puts on clean clothing, hanging out to air all clothing worn in sick-room before sending to laundry."[51]

Concern for neither the public's health nor the needs of these families could overcome suspicions of the germ-carrying capacity of the visiting nurse. For the most part, VNAs and boards of health were unwilling or unable to finance nursing service for the poor with contagious disease, choosing instead to allow them to do as best they could—home alone. Not surprisingly, then, only a small percentage of VNAs cared for patients with contagious diseases, at best delivering only "door step" instructions.[52]

"The Problem of Many Tongues"

Annual reports of VNAs included descriptions of the nationalities of patients, to illuminate the ethnic, cultural, religious, and economic differences between those served by the visiting nurses and their lady supporters. At the turn of the century about fifteen nationalities were listed, but a decade later as many as thirty were represented in many annual reports. The agenda to transform new arrivals into healthy Americans was apparent in the promising captions under pictures of patients whose "differences" were inescapable.[53]

With widespread acceptance of the germ theory of disease and an increasing realization that individual health depended to some extent on the health of the general population, the hazards of infectious diseases among the poor, especially immigrants, became an increasingly tangible concern. In the search for solutions to these urban threats, the visiting nurse seemed to provide a compelling answer to the complex problem of elevating poor immigrant families to a more ordered and healthier realm of well-being.[54]

Throughout the country, nurses cared for patients of many nationalities. An editorial compared the visiting nurse's task to the game of croquet in *Alice in Wonderland,* suggesting that "the whole thing was rather alive" and one had to regulate one's play accordingly. "Groups of different nationalities succeed each other in the same localities with a certain definite regularity— they seep and flow over various parts of the city as if obeying some tidal flow, the origin of which is no doubt connected with the opportunities for their own material betterment."[55]

Among the immigrant patients, growing numbers of Irish, Italians, and Russians caused special concern. The nurses could not understand their languages, and the immigrants' customs became familiar only as daily encounters brought these patients and their nurses together. The "natural ideas" of immigrants, those handed down from grandmother to mother to child, made the nurse's work inordinately complex. More comfortable relying on "superstitious observation and charms," patients often reacted to the nurse as a "most cruel and unfeeling monster" and responded to her hygienic message with hostility. Nurses often viewed Germans in a more positive light than other immigrants, describing them as thrifty or good.[56]

Perceptions of patients were frequently conveyed in sections called "nurse stories," an essential part of the VNAs' annual reports. Used, no doubt, to ensure ongoing financial support for the organization, these stories were consistent in theme and image. Patients were described as dependent, ignorant, improperly nourished, and occasionally shiftless and intemperate. Dwelling places were small, overcrowded, and poorly ventilated. One typical annual report noted homes where "the dirt of many years has accumulated and where the absence of cleanliness, while deplored, finds some reason for its non-existence." Immigrant patients were described as disregarding the most ordinary rules of hygiene, even during illness. When immigrants fell ill in "such unwholesome circumstances," their recovery was "either not to be hoped for, or at best very slow and unsatisfactory."[57] Not surprisingly, many nurse stories stressed the danger to the general population from immigrant patients with strange customs. Under headings such as "Protect Your Home," the stories exposed the ignorance of immigrants, the danger of the wants of their offspring, the misery of their diseases, and crime. These messages left little doubt as to the source of the problem or the solution—all that was required was more money to support the work of the visiting nurse.[58]

No matter how the nurses viewed their immigrant patients, the patients' habits were regarded as peculiar. Feeding pickles or pork chops to infants was seen as shocking, while practices such as wrapping infants in long bands "to keep them straight" or fears of opening windows at night were seen as simply "ignorant." In some instances, nurses described patients' behaviors as not only incomprehensible but life threatening. A story in the *Visiting Nurse Quarterly,* for example, told of parents who, having taken a sick baby to the doctor, failed to give either treatment or medicine because the child was "too much sick." Leaving the infant "untouched, uncared for, in a very dirty cradle," they had, to the dismay of the nurse, covered the baby with the warm blood of a chicken and then tied the bird's legs to a chair. This superstitious act was reported, so the story's author claimed, to demonstrate the immigrants' "great need of education." The teacher required under these circumstances was, of course, the visiting nurse.[59]

As nurses became more experienced in the problems of immigrants' Americanization, they tempered their message. To succeed, nurses increasingly were called upon to deliver "wise teachings," with tolerance, but the nurse's knowledge was often based on myth, stereotypes, and a limited appreciation of cultural backgrounds and different traditions of birth, marriage, and death. Over time the nurses gained familiarity with foreign languages and began to overcome their "ignorance of the immigrant's point of view." Some nurses used foreign language phrase books to enhance their vocabulary. "Winning her way" with immigrants was also aided by distribution of materials in patients' languages. In some VNAs, nurses formed clubs devoted to understanding the ethnically diverse community they served and discussing the "peculiar needs" of various nationalities.[60] VNAs were also eager to hire nurses of the same ethnic origins as their patients. The IDNA's motivation for wishing to hire an Italian nurse is clear in this description: "What we want in an Italian nurse, however, is more than merely a knowledge of the language. She could prevent many of our well-intentioned but often mistaken efforts." This strategy of ethnic understanding also helped fund the nurse's salary. For example, the Boston ladies were successful in convincing the Jewish Philanthropies to pay the salary of "a Jewish nurse who spoke Yiddish and could help us understand our Jewish patients." At the time (1913), the IDNA had seventeen hundred Jewish patients.[61]

The evolution of the immigrant into an "American" was measured in many ways, not the least of which was personal habits. Edna Foley, head of the Chicago VNA, mused, "Sometimes I think that nightgowns are a better test of our success in Americanization of our foreign born than about anything else that we give them, for a woman who has been taught not to sleep in her clothes and to undress her children at night before she puts them to bed has learned a great deal."[62] Wisely, the nurses also learned the importance of yielding to traditions that were not "permanently harmful, gaining friends instead of arousing resentment." It was acceptable to "leave the baby of Jewish parents unbathed on the day of circumcision . . . the Italian baby tightly swaddled until the opportune moment comes for urging it to be unwrapped!" Clearly unacceptable was the nurse who shouted at non-English-speaking immigrants to make her meaning clear, and the nurse who "would not demean herself by accepting hospitality from the foreign born."[63]

Whatever the conditions and circumstances of a patient's home, the nurse was rarely alone, for neighbors were always an integral part of the visit. Inevitably, explained Lillian Wald, "the flock of neighbors" would appear, drawn by "curiosity, idleness, interest, or sympathy." "Firmly dismissing the curious and idle," the nurse often found it helpful to accept the assistance of "any intelligent person" who might appear. Often forced to rely on a neighbor as interpreter, many nurses felt that much of the "enthusiasm, force, and personality" of their message was lost in the process. This was particularly true, claimed one nurse, if the interpreter happened to be a neighbor with eleven children of her own, "even if they had all died."[64]

The Unteachable Message of Health

By 1910, death rates, especially for infectious diseases, were declining, in part because of improvements in housing; water, food, and milk supplies; removal of garbage, ashes, and dirt; cleaning of streets; disposal of sewage; and control of contagious disease.[65] But freedom from disease no longer depended solely on community efforts or the construction of more public works and services. According to this new public health message, nothing was to be gained by uplifting the masses unless each individual uplifted himself, for "the effort of the few in trying to change the exterior conditions of the many is love's labor lost."[66] Winning the fight against disease required taking the next step from public to private hygiene. Hygiene was defined as the art of preserving health, of obtaining the most perfect action of body and mind. Its aim was to render growth more perfect, decay less rapid, life more vigorous, and death more remote. In a shifting paradigm of what constituted the public's health, the campaign to improve hygiene would focus on the affairs of the household and the conduct of the individual life. "What happens to the bottle of milk after delivery? How are the dishes washed in the kitchen sink? Who uses another person's towel in a common bath?" Answers to these questions were of greatest importance.[67]

If the public's health was increasingly focused on private matters, then the task ahead required translating the knowledge of scientific medicine into terms of personal effort and responsibility. C.-E. A. Winslow, a leading proponent of these new views on public health, declared that education was crucial if the average American's vast ignorance about health was to be turned around.[68] For the nurse, this meant using knowledge of the nature of disease, that "complex expression of the sum total of the interaction of parasite and host, a matter of relationship and relativity of many factors," to educate the patient.[69] The focus was not germs or even the cure of disease but knowledge of underlying causes. To see the problem otherwise was placing the cart before the horse.[70] Predisposing factors were those that lowered vitality or diminished physiological resistance: bad ventilation, poor and insufficient food or "ill-balanced" diet, exposure to heat and cold, insufficient exercise, an improper amount of sleep, too much or too little clothing, contaminated water or milk, and bad habits with regard to stimulants and other excesses. An explanation for this seemingly complex balance was found in the "laws of modern hygiene," which were reminiscent of the sanitary ideals of Florence Nightingale, though slightly modified by the germ theory of disease.[71]

The major elements of modern hygiene included sunshine, cleanliness, fresh air, and pure food. Sunshine was understood as particularly useful because recent studies suggested that it killed certain bacteria. It thus followed that dark rooms, shaded rooms, or rooms with northern exposure were "distinctly unhealthful and would depress both mental and vital activities, leaving an individual defenseless against disease." Simple observation of the disastrous effects of placing a healthy plant in the cellar for a few days seemed reason enough to accept the truth of these maxims. Light also had secondary benefits, encouraging cleanliness by "pitilessly" illuminating dust and other materials "inimical" to good health. Cleanliness was considered the best safeguard against disease. Elimination of dirt, dust, and flies, all carriers of disease, was the first condition of health, but cleanliness also extended to people, food, milk, and water.[72]

Clean, pure air and proper ventilation with life-giving oxygen were essential to good health. Rebreathing expired air with all its impurities undercut an individual's health. Impure air was as harmful to the lungs and general health as was tainted meat or spoiled fruit to the stomach. The teachings of Florence Nightingale remained in force: windows were made to open so as to flush the room with fresh air without causing a draft; doors and transoms were closed to keep used air from contaminating the rest of the house. Finally, a proper diet was required. To ensure good health, people were encouraged to replace cheap and indigestible foods, such as cabbage, turnips, doughnuts, and pies, with nourishing foods—fish, meat, eggs, rice, beans, hominy, and oatmeal.[73]

While believing that health in the home would ultimately mean health everywhere, visiting nurses also emphasized that the chain of health was only as strong as its weakest link. This created an endless set of responsibilities and opportunities for the nurse. Her first task was to enter the homes of the poor as a scientific investigator to uncover the moral as well as physical conditions of the whole family, "to root out the causes of illness." Having diagnosed a home's problems, the nurse then showed its inhabitants the way to health. The successful nurse could "impress her points upon others and make them see that what she proposes is right, reasonable, and advantageous."[74] Families often needed to be shown, as well as told, the elements of proper hygiene. Patience was an essential part of the nurse's equipment, for she "must be willing to reiterate over and over again, without showing annoyance, the rules which have been needlessly and exasperatingly ignored." No one knows better than the nurse, declared Ellen LaMotte, "the awful hiatus that exists between preaching and practicing—the glib promise and the broken pledge."[75]

Even as they acknowledged the opportunities created by this new endeavor, several nurse-leaders questioned whether the lives and homes of the poor provided the proper opportunity for successfully teaching the laws of hygiene. For the domain of the poor, as Isabel Lowman graphically described it, was at the end of a very narrow, dark alley where garbage and refuse covered the steps and ground with a kind of "loathsome litter," and where the floor was covered with banana peels, sticks, straw, dirt, and bits of food. As if the disorder were not discouraging enough, the "slatternly neighborhood woman was weaving endlessly, wandering back and forth and around the kitchen."[76] Families often lived in two-room tenements with the front room serving as a sitting room, kitchen, and laundry, usually with a sofa where at least one family member slept. The small room opening from it, with a little window at the back, held a larger bed. Water was carried from a landing in the entry. All discharges were taken to wooden privies in the yard, unless they were emptied into the sink. Typically, parents and several children would occupy very tiny spaces.[77]

For the poor, good hygiene was often a difficult, perhaps unreachable goal that "appears a mere mockery to those without material means of carrying it out." As Lillian Wald suggested toward the end of her career, if the message of health was to be carried to "troubled families," something also had to be done *for* those families. Other approaches might be sincere but were rarely realistic.[78] Other people would "bear effective witness" that only when the nurse came into the home and nursed the sick—did something when people were suffering—were families going to take her advice.[79] "Arousing the motives of self-preservation and self-interest, love of family, altruistic scruples against scattering infection, pride in the neighborhood or community, etc., which may ultimately assist in the formation of new habits if the physical difficulties are not overwhelming" remained the visiting nurse's greatest challenge.[80] At the same time, voices were raised suggesting that this was also the least appropriate agenda for the visiting nurse.

As Charlotte Aikens asked her audience at the 1906 National Conference of Charities and Corrections, "What can you hope to teach when you find these conditions?" What permanent results should be expected? What use is it to insist the patient should have plenty of nourishing food when "the whole family lives on baker's bread and a few green vegetables" cooked in oil, or to talk of cleanliness when a family of eight occupies one room and takes in boarders? Where the conditions "that make for cleanliness and virtue and self-respect are wanting," she concluded, there could be no real opportunity for educational work. Palliative advice could be given, but anything more, she suggested, was "useless tinkering."[81]

After several years of working with tuberculosis patients, Ellen LaMotte reluctantly admitted she shared these views. When she first entered the field, LaMotte naively believed that tuberculosis was both preventable and curable. Simple education about the nature of this disease and its transmissibility and spread would produce good results. Unfortunately, she found education impossible to put into practice because patients were poor and thus lacked will, intelligence, and self-control. For the same reasons, their environment was difficult to alter and cure was almost impossible. Eventually, LaMotte concluded that the nurse's teachings, no matter how thorough and conscientious, were simply partial measures; at best, she could hope for only momentary cooperation. In her 1908 paper "The Unteachable Consumptive," LaMotte reported findings based on her study of 1,160 patients, declaring that attempts to teach them either hygiene or prophylaxis failed no matter what the duration or effort.[82] In another study several years later, several nurses reported the effects of education among consumptives and concluded that, given six months or more of training, most patients were teachable. Of course, the "time and persistence" required to obtain these "good results" represented a rather substantial investment. Success meant that patients had learned to observe most or all of the following lessons: (1) care of sputum, (2) sleeping in separate beds, (3) precautions in coughing and sneezing, (4) not kissing children, (5) using separate dishes, (6) ventilation, and (7) cleanliness. Cost and benefit to the public of this type of work in poor patients' homes could easily be questioned. As LaMotte and others concluded, perhaps some patients should be treated in institutions.[83]

Many agreed with Mary Lent, Superintendent of the Baltimore Visiting Nurses, that education to suppress disease failed because the poor, even with help, were unable to "consistently and unflaggingly" apply what they had learned to their daily lives. Very poor families simply had no means for carrying out the principles of hygiene, nor did their surroundings permit it.[84] Nurses who had been in the field the longest were now willing to admit the truth. Lent declared that, at best,

> we are but Red Cross nurses in the field of battle. For years we have been giving temporary relief in the way of skilled nursing care, under conditions that deprive it of three quarters of its value. We have also been trying to teach underpaid, overworked, underfed, wretched human beings how to live more hygienic lives. But the awakening has come at last. Our eyes are now open to the facts. We can no longer continue to dole out surface relief and believe it stands for anything more radical. It is for us, who are palliative agents, to declare that the conditions of today do not call for palliative treatment.[85]

Lent concluded that the nurse's most valuable work was the collection of facts about conditions that demanded nursing care while simultaneously defeating nursing's efforts. The duty of nursing was to present these facts to the public "in such an array and in such numbers that they can neither be contradicted nor ignored."[86] Many other visiting nurses shared Lent's view that the problems of the poor required social reform more than hygienic instruction. Lillian Wald went so far as to suggest that impressing upon the poor the last word of science without simultaneously urging reform in housing, child protection, and wages was "cruelly sardonic on the part of the nurse."[87] Of course, there was always the unthinkable possibility that hygienic living had nothing to do with preventing the spread of tuberculosis or anything else!

During these early years, visiting nursing was an experiment in caring for the sick at home. Determining which patients could be cared for successfully at home, providing essential family support, and overcoming the impediments to caregiving caused by customs, language, and poverty were all-consuming tasks. As visiting nurses and their lady managers grappled with the complex matter of defining a legitimate and appropriate role in the community, they increasingly confronted and understood what agendas could be pursued. Weighing the costs and benefits of home care raised the inevitable question of whether all patients required the skilled services of a nurse or whether some—such as the chronically ill—could be cared for by a less highly trained attendant.

From any perspective, home nursing was a challenge. For the poor, living in crowded tenements and confronted by dangerous illnesses, the visiting nurse was one of the few real sources of assistance.

CHAPTER 4

Caring in Its Proper Place

Race Relations at Home

*C*aring for the sick at home, even under the best of circumstances, is a complicated undertaking. Patients and their home lives are resistant to edicts of all sorts, including those created to separate the races. Not surprisingly, the management of race relations in patients' homes has varied from place to place and over time. Charleston and Philadelphia are presented here as representative of a small southern town and a large northern city in which we can see the many critical elements of caring "in its proper place," with its delicate balances and boundaries between providers and recipients, black and white.

These case studies illustrate what happened when race relations were challenged by the practical concerns of home care. The Ladies Benevolent Society of Charleston exemplifies the one-nurse agencies that provided most (68%) of the care for the sick at home. In contrast, Philadelphia's Visiting Nurse Society is representative of the large organizations that dominated developments in home care yet constituted only 3 percent of the agencies delivering such care. In both cities, race, class, and gender remained forceful determinants of who received and who delivered care at home, as well as the quality and outcomes of that care. Charleston and Philadelphia illustrate, in the words of historian Glenda Elizabeth Gilmore, the "stiff-sided box of place" dictating interactions and defining boundaries for locally acceptable caregiving practices between the races.[1]

Racism and sense of place were not purely southern sensibilities. But these convictions were more obvious in the South and references to them are therefore more evident in southern records and accounts.[2] Although patients mostly remained the silent objects of care, their stories, as recorded by others, were similar. Especially in Charleston, the written record of white lady managers and black nurses documents the impact of race on care at home. In northern cities, however, racism was less socially acceptable and therefore documented less prominently. The comparative silence of white northern lady managers and visiting nurse records on the "race question" is impressive. In Philadelphia, records rarely mention race and annual reports simply record numbers of black patients in the caseload. Throughout the North, it was simply assumed that black visiting nurses cared for black patients only and white visiting nurses saw both black and white patients.

Only a few studies, a rare journal article, and an occasional letter from a leader in visiting nursing mention race. Northern white attitudes toward black colleagues and patients are asserted in texts (written by whites) intended to promote the betterment of the black race. The writings of black nurses clearly document their pleas for a chance to do more for the black community. Mixed voices, white and black, reveal telling aspects of the impact of racism on the black recipients of home care and their caregivers. Their stories remind us that the private, intimate, and often invisible nature of home care mandates respectful, trusting, and competent relationships between providers and recipients of care—no matter their race, class, gender, or ethnic origins.

Charleston: A Southern Exposure, 1880–1925

The Civil War and Reconstruction left Charleston in ruins, its civic institutions described by one contemporary observer as "utter topsy turvy." The city was a mecca for freedmen, whose numbers quickly exceeded those of the white population, bringing both freedom and rising racial animosity, unemployment, crowding, and communicable diseases.[3]

By the 1880s, however, wealthy Charlestonians had resumed their former lifestyle. Race and class contributed to the domination of social, political, and economic life by the same few established families. While maintaining a certain tradition that outwardly stressed tolerance and civility, white Charlestonians, like most of their countrymen, quickened the pace of racial discrimination and enforced institutional separation. Troubled by the persistent presence of the destitute poor and hoping to replace "willy-nilly" impulse charity with a more modern approach, affluent citizens organized the Associated Charities Society of Charleston in 1888. Borrowing ideas from older charity societies in the North, Charlestonians implemented a system

of scientific charity and hoped to reduce by half the number of persons requiring assistance.[4] Early on, the society turned away many applicants, re-
sponding primarily to native white families of unskilled workers. Black applicants were rare, no doubt discouraged by the society's regulations and
requisite investigation.[5]

Caring for fourteen hundred patients (black and white) annually, the municipal hospital was the largest dispenser of charity care in the city during
the 1880s. Financed by city appropriations, "attendance upon the sick and
the general order of the hospital" were provided free of charge by the Medical College in exchange for the school's use of the hospital for clinical purposes. By 1882, the hospital board of commissioners, "always solicitous for
the welfare of the indigent sick," sought some practical way to provide educated nurses to meet the needs not only of the public hospital but also of the
"family circles of the rich." An act of the South Carolina General Assembly
launched a training school for white nurses in Charleston. Three graduate
nurses from "elsewhere" (New York Hospital Training School) initiated this
enterprise, which after several false starts emerged as the Roper Hospital
Training School for Nurses. By 1905, the school's graduates numbered fifty-
two.[6]

As in black communities throughout the country, black Charlestonians
organized to meet their common needs. By 1897 they had established the
Colored Relief Society, but for most the black church remained the central
source of social welfare. In 1896, supported by the city's notable black business and religious leaders, a group of six black physicians and dentists
opened the Hospital and Training School for Nurses in Charleston. Founded
in response to the denial of staff privileges to black physicians and the exclusion of black women from admission to the City Hospital Training School,
the hospital survived until the late 1950s. According to historian Darlene
Clark Hine, it was one of the more resilient black health care educational institutions in the South.[7] Through the creation of their own hospital, Charleston's black physicians and nurses made a place for themselves—though
admittedly separate and never equal. The hospital was seen as the black
community's response to its needs for health care. Although the hospital and
school were created, sustained, and controlled by the black community, their
advisory board was composed of both black and white citizens.[8]

"OLD METHODS CAN NOT RUN A NEW WORLD"

Like the city's other charitable health care institutions, the Ladies Benevolent Society (LBS), through its white lady managers, also reclaimed a tradi-

tional concern for care of the sick poor. Membership and income in the LBS remained small, however. The committed few found it strange that an effort "which appeals to the tenderest sympathies of our nature should not attract more interest." The small scale of the enterprise was attributed to "the numerous charitable societies in Charleston, the many private calls for aid to which all seem subject," and the reality that many older members' "fortunes were lost in the vicissitudes" of their times.[9] Catherine Ravenel, elected superintendent of the LBS in 1895, later recalled feeling disheartened as, year after year, the relief given was never adequate to the needs. Looking forward with anxiety, Ravenel noted that "old methods can not run a new world."[10]

In 1902, Ravenel read a copy of Harriet Fulmer's article "The History of Visiting Nursing Work in America." She realized immediately that, while the LBS was the oldest home care organization of its kind in the United States, organizations in fifty-three other cities were using trained nurses rather than lady visitors to care for the sick at home. Ravenel concluded that the trained nurse was the best means at the smallest cost to help the poor.[11]

At first the LBS board did not share Ravenel's enthusiasm for trained nurses, but it agreed to attempt the undertaking for a few months. For Ravenel, hiring the first trained nurse lifted a heavy burden, giving the "old society" new opportunity and inspiration. Following Fulmer's advice and "borrowing" heavily from her rhetoric, Ravenel appealed to the public for understanding and support. In the newspapers, she proclaimed the virtues of trained nurses and home care, declaring theirs "a hand to hand fight against disease, poverty and dirt, the most pitiful ignorance and inherent prejudice."[12] The trained nurse could now take the advantages of the hospital into the homes of the poor, saving them from the anguish of sickness and death. As Ravenel explained, trained nurses were far more efficient than the nurses of old times. Thanks to careful training and strict requirements for respectability and character, the profession of nursing was "elevated to so high a plane that no lady need hesitate to enter it; to say nothing of its being one of the noblest and most womanly occupations."[13]

Despite her eloquence, Ravenel later admitted she had taken on a very hard task. Not only did she need to convince a very tradition-bound board to reorganize the work of the LBS, but the doctors were "skeptical" and gave the board no support. Even the poor did not understand the advantages of a trained nurse. According to Ravenel, "We forced it upon them—we hunted patients and interviewed the doctors." In any event, over the next few years, the number of physician (black and white) referrals and patients grew steadily. On average, 250 cases were helped each year; 30 percent received friendly visits from the ladies, 5 percent were paying patients, and 65 percent received free care from the trained nurse. Beyond caring for the sick, the LBS found a growing number of maternity patients seeking its services. Responsibility for accepting referrals for care and investigating all doubtful cases was assumed by the Associated Charities Society, which also provided the LBS with a room and two "commodious" closets for supplies.[14]

ANNA DECOSTA BANKS: "SHE HAS DONE WHAT SHE COULD"

In a 1903 newspaper article, Catherine Ravenel explained how trained nurses had "a fascination" with caring for the sick at home. From the outset, Ravenel was committed to providing the LBS with a "nurse of tact, executive ability, and one who can command respect and win love, yet is willing to do the most menial service, one having that indescribable something which brings comfort and rest where all is suffering and darkness." As might be expected, the LBS spent three years finding such a nurse. Much to the society's dismay, the white nurses of Charleston were unable to survive the demands of such work, succumbing to illness or better offers within weeks. During a very short period, five nurses were employed and then left the job. Finally, in 1903, the LBS hired a black nurse, Anna DeCosta Banks, in "desperation" and only temporarily, to fill in between two white nurses (Figure 9). The society decided to make a one-year commitment to her in 1906 when, in the middle of an epidemic, a white nurse abandoned her job and Banks once again came to the rescue.[15]

In an annual report, Banks was described as "a most uncommon nurse—everything that one could desire." She was "judicious, tactful, and experienced" and "so skillful, capable, and tender-hearted." She was never sick and unable to work, was "very constant in her visiting," and, unlike one of her predecessors, did not abandon the society during an epidemic. Notwithstanding these praises, because Banks was black the ladies only reluctantly employed her for the long term. The Nurse Committee reminded the board that it had been very disappointed in the efforts to secure the services of a white nurse, and if a white nurse was unsuited to the work, surely a black nurse would prove no better.[16]

Anna DeCosta Banks, however, was an exceptional nurse. Born in Charleston in 1869 and educated in Charleston schools and at the Hampton Institute in Virginia, she graduated in the first class at Dixie Hospital School of Nursing at Hampton. Banks later recounted that her whole life had been shaped by the training at "dear old Hampton." Having received special funds

FIGURE 9. Anna DeCosta Banks. *Waring Historical Library, Medical University of South Carolina.*

to attend Hampton, Banks spent her life trying "to do for others what [was] done for me." In 1912, she wrote that this institution and its leaders gave her inspiration and strength whenever she was tired or discouraged.[17] Banks's description of her motivations nicely illustrates what Hine describes as a mutual dependence between black nurses and their community—a kind of belonging in which the black nurse labored within and on behalf of the community.[18]

Returning to Charleston, Banks became head nurse of the Hospital and Training School for Nurses (black) when it opened in 1896, and was later appointed superintendent.[19] Typically, the hospital's work was among the black community's most destitute, "people who are without homes, money, and friends." The hospital had two wards containing four beds, three private rooms, an operating room, and a dispensary. Eight months after the hospital opened, 135 patients had been cared for in the building along with forty-two "outside cases." Only half of the patients were able to pay the $3 weekly fee; patients more often gave potatoes, rice, corn, eggs, or chickens in place of money. The hospital's main source of income was "what we get from our nurses at private cases." As in many hospitals, these private nurses were actually students sent to work in white and black homes, with all fees paid directly to the hospital. Banks saw this as an opportunity for students to work with white physicians in the homes of private patients of means, as well as a chance to ensure that colored women's traditional role as caretakers of the sick did not "drift into the hands of white nurses." Banks believed that "black girls had to be trained to take their places" and "patients and their physicians (white and black) required the opportunity to learn that they could rely on black nurses for care when sick." By 1905 she concluded that "the South [was] the place for the colored nurse, there [was] plenty to do and they were well appreciated."[20]

In reality, the practice of sending student nurses into patients' homes provided few learning opportunities, since they often missed classes and supervision was minimal. In her autobiography, Jane Edna Hunter illustrated the challenges she faced as one of these students. Hunter, who would later found Cleveland's Phillis Wheatley Association (a residential and job training center for black women), claimed that after only six months of training she suddenly became "Nurse Hunter" when sent to care for a child sick with scarlet fever. Her first case was with Charleston's influential Rutledge family in their elegant home on the South Battery. Hunter described working hard to save the patient and her dignity, fearing that her "whole future might turn on how she handled it."[21] Ironically, this practice, common in white and black nursing schools, was called "hiring-out," a term used by slave holders when temporarily renting a slave's services to others and personally collecting the income.[22]

To support the Hospital and Training School, Banks depended on funds and donations from the community. She described potential black donors as of two types: "those people who can do [but] will not, and those who would do more if they could." Her fundraising efforts often extended to the white community, where she identified generous friends who gave window screens, interior whitewashing, and a bed or two. As an active member of Charleston's Young Women's Christian Association, founded in 1907 for the benefit of improving conditions among the "colored people," she was involved in securing homes for working girls and gave lectures demonstrating nursing techniques.[23]

Given her impressive background and knowledge of race relations in the city, it is understandable that Banks had reservations about working with the LBS. Expected to care for both races and knowing the feelings in Charleston "between poor whites and the Negroes," she agreed to "try it" for a month. The ladies, described by Banks as of the better class of white people of this city, were "so much pleased," and she found visiting nursing enjoyable. Admitting that care of ignorant white people required a "lot of tact," Banks quickly found that when people were sick or in need "it does not make any difference to them who you are or what color if you have come to help them, all are gladly received." In Charleston, she would later claim, "our real homefolk understand each other perfectly." When trouble between the races occurred, it was, according to Banks, brought by outsiders.[24]

The decision of the LBS to hire Banks did not go unnoticed or unchallenged. The most severe criticism was voiced in the *Nurses Quarterly* by the white nurses of Charleston. Such criticism, however, did nothing more than provoke a great deal of discussion and an admission by the LBS board that it preferred a white nurse, if it could find one who "would keep up the work steadily."[25] Catherine Ravenel sought the Medical Society's "opinion as to the success of the work and if they [had] found it of value to them in their profession." The Medical Society wrote a letter expressing "unqualified" support of the LBS's work and of Banks, and this was a "great encouragement" to Ravenel. Confident that the physicians did not share the white nurses' concerns, and undoubtedly aware that these white nurses also needed to remain in the good graces of the Medical Society, Ravenel tried to avoid further criticism by publishing the Medical Society's letter.[26]

Ravenel's enthusiasm for Banks still was not shared by several board members, and for years the annual rehiring of Banks precipitated debate among the members over "the question of the Society's opportunity and ability to secure a white nurse for its visiting nurse work." While Ravenel spoke "at length of the advantages of a colored nurse" and the board freely "debated" the matter, no alternative was found. The members of the "White Nurses Association," who had criticized the hiring of a black nurse by the LBS, could offer help only for an hour or two in emergencies. Hoping the

LBS could train its own nurses, Ravenel contacted Roper Hospital's Training School seeking a pupil to "occasionally visit with the nurse to learn this
work," but "none could be spared." She even spoke to Ella Crandall, the executive secretary of the National Organization for Public Health Nursing,
about "her opinion of colored nurses in New York," where nearly half of all
black public health nurses in the country worked. No matter how divided
its opinion, the board continued to rehire Banks until she died, in 1930.[27]

"OPEN THE DOOR OF YOUR HEART, MY FRIEND, HEEDLESS OF CLASS OR CREED"

Through her work with the black hospital, the training school for nurses,
and the LBS, Anna DeCosta Banks hoped to create an inventive matrix of
care for the black community and simultaneously accustom the white community to care by black trained nurses.[28] She pursued this goal by offering
her students as a supply of extra nurses. With no need to distinguish between
pupils and trained nurses, the LBS saw these women as a flexible and affordable solution to the demands created by a fluctuating caseload. The services of the black pupil nurses were accepted with confidence, for the school
was held in high regard and maintained a system of discipline described as
severe but necessary. Not only did Banks have access to an easy supply of
"nurses," but she made it possible for the LBS to pay $1.00 per day, rather
than the usual rate of $10.50 per week, and to receive twenty-four hours of
free care in emergency cases. While the work of the LBS created an environment of mutual benevolence for the pupils, patients, physicians, and
ladies, it also provided the Hospital and Training School with a dependable
source of income essential for its survival.[29]

Although the LBS had cared for black patients for at least seventy-five
years, it did so in a carefully controlled fashion.[30] Predictably, Banks's initial
efforts to extend the society's services to a wider black community proceeded quietly, acknowledged simply as Banks's occasional visits to "respectable old Negro house servants . . . generally recommended by white
people." By 1910, however, a new section, "colored patients," appeared in LBS
reports, documenting over time the society's evolving relationships with
black patients and their black physicians. Banks would eventually note in a
report that no person had been refused help. Ravenel's editorial comment
on this, hand-written in the margin, was "who came to me of course, there
are many others who need help." Clearly, Ravenel was in charge, but for
Ravenel and Banks these early years were filled with testing the allowable

boundaries between the LBS's mission and Charleston's particular definitions of race and class.[31]

Polite differences were expressed privately, and Ravenel respected Banks's
professional judgment and suggestions, including the creation of services to
meet specific needs of the black community. A delicate balance was required,
but, as Banks wrote to Ravenel, "I am as happy to be able to remain with you
during the coming year of 1908 as you are to me." Banks added that she
hoped it would be said of her "she has done what she could"—and indeed
it was. When she died in 1930 at the age of sixty-one, Banks was honored for
untiring service to her people. Her alma mater claimed that no other Hampton graduate had had a more vital influence on her community than Anna
Banks. "There [was] scarcely a colored home in Charleston which [had] not
felt the influence of this good woman."[32]

METROPOLITAN LIFE INSURANCE COMPANY

As the LBS board and Anna Banks struggled to determine the boundaries of
locally acceptable caregiving between the races, the greatest boost to the
black community's access to nursing care came unexpectedly in 1911, when
the Metropolitan Life Insurance Company (MLI) extended its visiting nurse
service across the country. Hoping to reduce the number of death benefits
claimed, the MLI contracted with local visiting nurse organizations to provide nursing care to policyholders during illness. Since 12 percent of the
MLI's policyholders were black, this meant that over one million black policyholders could now seek the services of a visiting nurse when sick at home
or pregnant. Forty-one percent of black policyholders lived in the southern
or southwestern states; lacking alternative sources of care, they took advantage of the visiting nurse service more than did white policyholders. Because
mortality rates for all MLI's southern policyholders were higher than those
for MLI policyholders in general, and because death rates of black policyholders were 50 percent higher than those of white policyholders, the work
of Metropolitan's seventy-eight new nursing services in 139 southern towns
was of particular economic significance. A decade later, the number of black
policyholders had doubled.[33]

Contracting with the MLI meant accepting patients without discrimination, and this confronted white lady managers across the South with the
need to rethink race relations within the very intimate context of the patient's home. Would they hire a white nurse to care for white and black patients, hire a black nurse to care for black and white patients, or segregate

patients and their caregivers by race? What if a black patient had a black physician—could a white nurse be "the right hand" of a black physician? Finally, if, as was often the case, the organization could afford only one nurse, would she be black or white?

Although Jim Crow laws dictated race relations, response to the "colored problem" took place within local contexts of acceptable behavior. Laws governing nurse, physician, and patient relations after the turn of the century were usually written for specific states and institutions, but private practices reflected local approaches to segregation and to the particular relationships created by caring for the sick. For example, only South Carolina and Mississippi law required general segregation in hospitals. These two states also required black nurses for black patients. Alabama law prohibited white female nurses from attending black male patients.

The home-based care provided through MLI contracts to black patients introduced yet another complexity in these attempts by white southerners to separate the races. Responses to this new challenge, expressed in the language typical of the period, varied from place to place. In Jacksonville, the sentiment was that "white people should not be caring for colored patients." A white nurse from North Carolina declared, "The Negro is our responsibility . . . a Negro knows how to respect a woman when she comes to the door. They are docile, they are grateful, they make excellent patients—so long as you have the ability to control them." In Birmingham and Nashville, white nurses characterized their black co-workers as "indispensable" providers of "necessary bedside care involving physical contact" in black homes, "a service the white nurse preferred not to undertake." In Richmond, where the "Metropolitan work opened [their] eyes" to the need for nursing care in the black community, the ladies found it "eminently" satisfactory for white nurses to care for black patients on the "same footing" as white patients—but drew the line at white nurses "taking orders from the colored physicians." According to their chief nurse, the best solution was to "put on a colored nurse who takes the work from the colored physician" and to seek the funds for her salary from the black community.[34]

The MLI first contacted the LBS in June 1911. By the end of a month of "testing the plan and learning the duties of the nurse," the ladies were confident that "the work will certainly extend and we trust our income will be increased." Agreeing to provide nursing service to the six thousand MLI policyholders in Charleston, they admitted "it [is] a venture for both the Society and the Company and we trust it will prove a great thing for us." Because there was no extra charge or reduction in benefits for MLI policyholders, Catherine Ravenel (correctly) believed the MLI's motivation was simply to use this service as a form of advertising. The objective of Metropolitan Life, she explained to the LBS membership, was to keep people alive—"as long as the policyholder lives, he pays the Company weekly installments. When the patient dies, the Company pays the amount of the policy. All's fair." After only six months of experience with Metropolitan, she also felt compelled to mention to her constituency that "this arrangement [has] increased our work, especially among the Negroes, as the nurse visits indiscriminately."[35] Similarly, Anna Banks described the impact of the MLI as "broadening our work and bringing us in contact with more colored people." It was "a blessing to the colored people especially, because they believe in the Insurance Company and I have a chance to get into their homes more."[36]

As Ravenel had hoped, the most immediate boost to the finances of the LBS came from Metropolitan payments for visits to patients formerly attended free. A decade later, the company was contributing 50 to 60 percent of the LBS's annual income, making it possible to hire a second nurse to keep up with increasing numbers of patients (from 173 to 887). This time the LBS board hired Cecelia Trescott, a graduate of Harlem Hospital, to assist Banks.[37] While the ladies still promoted interest, enlightenment, and warm hearts, hoping to "open the purse" for the LBS, little else remained the same. Visiting nursing was suddenly more complicated, requiring a new system of record keeping, billing, and cost accounting. The patient's status with the MLI had to be verified so that care to "undesirable policyholders" would be immediately noted. Visits were regulated by the nurse in consultation with the patient's physician, and great emphasis was placed on visiting as soon as possible. Most dramatic was the continuing change in the racial composition of the patients, with blacks increasing from 20 to 45 percent of the caseload. In a gesture calculated to minimize the LBS board's anxiety about growing involvement with the black community, Ravenel's annual report for 1920 concluded that, although "the Society did not [initially] contemplate helping Negroes—now, to a limited extent, this is imperative."[38]

The LBS and Metropolitan maintained a mutually satisfactory relationship for many years. According to Ravenel, "the Company even claimed that from their [MLI] efforts and ours [LBS], the death rate among their policyholders had been greatly reduced in Charleston and they were very proud of it." Despite the MLI's "complete satisfaction" with the work of the LBS, in June 1921 rumors circulated that the company's support was to be given to another organization. Two years later, the LBS was visited by a Metropolitan field nursing supervisor from New York and the associate manager of the Charleston office, who claimed that "our policyholders did not call on the nurses as much as Metropolitan desired." The problem, they claimed, was "that [policyholders] did not like colored nurses and feel they are acting as spies." After spending several days in Charleston investigating the situation, the two Metropolitan visitors recommended closing the contract with the LBS.[39]

These complaints about black nurses came primarily from "group" policyholders afraid that the nurse might know too much. Group policyholders were a new market initiated by the MLI in about 1918. Group policies offered companies a broader range of services than those available to individual policyholders. Under such agreements, when companies "had good reason to believe" an employee was absent from work because of illness, the MLI provided routine nursing services to the worker and then sent a nursing report describing the employee's condition and probable duration of illness. If the nurse found a serious social problem, she was expected to share this information with interested employers in a confidential report, but the MLI claimed the nurse's role was not to investigate and the reports were provided simply as valuable aids to companies in planning production. Such reports no doubt were the source of racially motivated paranoia by group policyholders. Much more difficult to explain were the MLI's concerns about rate of usage of the visiting nurse, since the average number of visits to policyholders (200) per month was clearly within customary standards for visits.[40]

On 3 February 1923, Lee Frankel, Metropolitan's vice president, wrote to Catherine Ravenel wishing to "bring before you the facts that are being brought before us daily." Expressing his appreciation for the good work done by LBS nurses, he wrote that the company was obligated to withdraw its contract because of policyholders' objections "to nursing service being given by the colored nurses." He trusted that Ravenel would "appreciate fully our stand on the matter" and Metropolitan's need to give policyholders the "type of service they demand." The LBS board immediately acknowledged Frankel's cancellation and began plans for "the necessity of reducing the expenses of the Society." The board dismissed the second nurse (Cecelia Trescott), sending her "a testimonial of appreciation," reduced its office expenses, and notified the city's physicians that, with termination of the Metropolitan contract, the society would "be better able to answer all calls" and to care for either destitute or paying patients.[41]

The recorded events fail to explain the LBS's apparent acquiescence to termination of this mutually beneficial relationship. Ravenel merely claimed that what the society received did not cover its output, and it could not afford to engage more costly white nurses. She reminded the membership that the board had tried to find white nurses but had learned that "success in this work depends on the character of the nurse." Apparently finding white nurses somehow lacking in the requisite abilities, she concluded by simply stating, "Anna Banks is still our nurse." For the ladies of Charleston, Banks had become, at any price, the only acceptable bridge between the classes and the masses, white or black.[42]

As the secretary of the board explained several years later, the LBS did not want to give up the nurse who had served the society faithfully for so many years, and "the Society found that the colored nurses were more satisfactory as we send our nurses to either whites or Negroes, and of course, many times they have to nurse a case which is attended by a Negro doctor."[43] In other words, having extended the services of the society to the black community, the ladies found it unacceptable to hire white nurses to work with black doctors in the homes of black patients; black nurses could cross these racial, class, and gender barriers, but white nurses could not. It was unimaginable that white nurses should "serve" as the right hand of black (male) physicians or provide very personal care to black (male) patients. Under such improper circumstances, white propriety and white power were compromised.[44] Interestingly, solutions satisfactory to white ladies were roundly rejected by white working-class patients. What remains a mystery is why the LBS did not follow the familiar Richmond model of hiring a white nurse to work with the white community and a black nurse to work with the black community. Whatever the explanation, the correct relationship between patients and caregivers varied not only with race, class, and gender, but also with place.

By refusing to comply with Metropolitan's terms (to hire a white nurse), the ladies chose a very different future for their organization. By 1925, Ravenel would describe the work of the LBS as restricted. Its calls now came from doctors, established patients, and others who "preferred" its work. The society continued to serve the black and white communities, and when Anna Banks retired it "hired the colored nurse Claudia Green Adams" to replace her. As it had a century before, the LBS cared for an average of 250 patients annually in the years prior to the Depression. Ravenel contended the society still had "as much as we could conveniently handle."[45]

The LBS story allows us to examine the history of philanthropic care of the sick at home within a very complex social structure in which race, agreeable manners, "firm principles and compact prejudices" intermingled with a long tradition of benevolence.[46] For the ladies of Charleston, defining worthy recipients of care, rationing scarce resources, competing for the public's support, debating their appropriate mission, and preserving their institution intact constituted a longstanding agenda best explained by a shared appreciation of caring in its "proper place." Understanding Charleston's traditions, the LBS both engaged in battles to defend its agendas from distortion by outsiders and deferred only to the judgment of those who demonstrated an understanding of its worldview. For these ladies, appropriate methods of caring and spheres of usefulness were intimate matters defined at home within locally acceptable boundaries, and certainly not by strangers.

The Philadelphia Story: A Northern Exposure

At first glance, Philadelphia's history of race relations appears more capricious than Charleston's. Philadelphia was known for its participation in the movement to abolish slavery, in the Underground Railroad, and in the founding of the first antislavery society. Its black population experienced periods of progress and vibrancy, racial proscription and inequality, violence and race riots, and complex social tensions created by industrialization, urbanization, and immigration.[47]

As late as 1900, Philadelphia had more black residents than any city north of the Mason-Dixon line. The 1900 census counted 62,613 black Philadelphians, about 4.8 percent of the total population. Although the black population was large, it was far less visible than in Charleston. Black people in Philadelphia were never completely segregated; no ward was solidly black or white. Most black families lived in the city's geographic center. Attitudes toward the black community were mixed, with whites denouncing the excesses of southern racism while tolerating less obvious forms of prejudice in their own city. Whites and blacks did not intermingle comfortably with one another.[48]

PHILADELPHIA AND THE GREAT MIGRATION

By the 1920s, Philadelphia was experiencing a racial transformation caused in part by the great northward migration of southern blacks in search of a better life. The trend was initiated by northern demand for labor during

World War I and the boll weevil epidemic in southern cotton fields during those same years, and was sustained by the postwar industrial boom and the end of immigration from southern and eastern Europe. The peak years of migration were between 1922 and 1924, when more than ten thousand black southerners moved to Philadelphia each year. By 1930, with a net increase of just over eighty-five thousand, the city's black population reached 219,599, or 11.3 percent of the overall population.[49]

These southern migrants came from diverse backgrounds and brought different levels of preparation for northern urban life. As their numbers increased, so did the movement toward discriminatory employment practices and segregated schools, restaurants, theaters, health care institutions, and churches. Intraracial conflicts between old and new black residents added to these social tensions. In Philadelphia, the contempt of established blacks for these newcomers reflected their need to explain a newly growing racism and loss of social privileges. In January 1924, the Pennsylvania State Department of Welfare sponsored a conference of social agencies to initiate programs for "remedying the needs of its Negro population." The director of the Department of Public Welfare, Dr. Ellen C. Porter, reminded the interracial audience that "*Our* problem is not altogether with the Negro from the South. It is with ourselves."[50]

Following the conference, the Department of Welfare initiated a statewide survey of "the relationship between the Negro and his environment," directed by Forrester B. Washington, the black executive secretary of the Armstrong Association of Philadelphia (the local affiliate of the Urban League). The survey identified housing as one of the most serious problems confronting the black community. Although residential segregation did not legally exist in the state, restrictive covenants forced blacks into already overcrowded "sections set aside for the race." Most lived in homes located in small courtyards or alleys, where four or five families reportedly shared an outdoor hydrant and a common toilet. Taking advantage of this crisis in black housing, landlords charged such high rent that 80 percent of black families in Pennsylvania took in boarders. And bad housing, of course, was directly related to ill health. As Washington pointed out, "Disease cannot be segregated. It soon spreads from narrow alleys and side streets where it most often originates to broad paved avenues and boulevards."[51]

Other members of the black community acknowledged the perceived menace posed by southern migrants. As Mr. L. B. Moore expressed at the 1924 Inter-Racial Conference, "The colored people from the South have brought with them everything they have—money, household goods, small-pox—everything they have."[52] Such fears were reinforced by newspaper ac-counts describing "a virtual smallpox epidemic" among recently arrived blacks, resulting in the quarantining of hundreds of city blocks and vacci-nation of thousands of people. Smallpox, of course, was not brought from the South, but it developed quickly as unvaccinated black southerners were exposed to smallpox in their crowded neighborhoods. Coming to the urban North was dangerous and, for many, resulted only in further impoverish-ment, overcrowding, exposure to new diseases, and even death.[53]

White Philadelphians worried that close contact with poor blacks living in unsanitary neighborhoods was dangerous. In particular, upper-class white employers of black domestic service workers feared exposure to tu-berculosis. Occupational studies suggesting that domestic workers had by far the highest incidence of tuberculosis, and the claim by white physician Henry Landis of the Henry Phipps Institute that domestic workers were a constant source of infection, served to intensify white anxieties. In 1915, the all-white board of the newly organized Whittier Center declared the pre-vention and arrest of tuberculosis in the black community a top priority. As one of the nation's leading clinicians and researchers in the field of tuber-culosis, Landis championed many partnerships aimed at conquering tuber-culosis in the black population, including those with the Whittier Center, the Philadelphia Health Council and Tuberculosis Committee, and the Henry Phipps Institute.[54]

"A BETTER CHANCE FOR LIFE"

Health care for the black community came from numerous clinics, hospi-tals, and voluntary organizations.[55] Care was carefully cloaked, however, in a web of discrimination. A telling description appeared in a report by Ethel Johns on the status of black women in nursing, prepared for the Rockefeller Foundation in 1925. As part of her investigation, Johns interviewed Forrester Washington, considering him a primary informant on the Philadelphia situation. Having "probed and dissected" the status of race relations in Philadelphia, she concluded:

> The reaction of white people to the Negro situation is interesting. Not for nothing are they "on the Mason-Dixon line." Segregation is "not permitted" and yet tacitly exists. Most ingenious methods have been devised for separat-ing whites from blacks in hospital service without making the situation too ob-vious. The sharp increase in population with its implied threat of economic

competition gives rise to uneasiness if not alarm. The growing power of Negroes in municipal politics constitutes a problem for heads of civic departments including the health services. In short, Mr. Washington is right when he
anticipates increased tension between the whites and the Negroes.[56]

Other studies came to similar conclusions. Some hospitals limited the number of beds available for blacks, and others did not admit blacks; most
treated blacks in their outpatient departments. Often, hospitals had separate
clinics or clinic hours for black patients. Some hospitals admitting black patients nevertheless refused black maternity patients, who were encouraged
to have home deliveries assisted by medical students in need of experience.
Black patients were discharged early when ward beds were in tight supply.
Hospital care by white nurses varied from impartiality to patent discrimination. Under these circumstances, health care for blacks was less than satisfactory, and patients learned to avoid those places with a reputation for
discrimination.[57]

Most mainstream medical institutions refused to desegregate training,
staffing, or treatment practices, and this led to growing interest in black-
controlled hospitals and educational programs. By 1906, Philadelphia had
established two black hospitals with nursing schools within a four-block ra-
dius, making Philadelphia one of three important northern training centers
for black nurses at the turn of the century. The struggles of these institutions
with one another, and simply to survive, have been well documented. Black
graduate nurses from Philadelphia training schools typically found employment in private duty, working for both black and white families. Like
Charleston's black hospital, Mercy and Frederick Douglass Hospitals in
Philadelphia received support from both the white and the black communities.[58]

In addition to the care provided by Mercy and Douglass hospitals, blacks
obtained care from Philadelphia's Visiting Nurse Society (VNSP), the Department of Health, and specialized programs such as the tuberculosis programs run by the Henry Phipps Institute (Phipps). By 1925, thirty-eight
black nurses were employed by these five organizations. The city was also
served by 165 black physicians, sixty-nine black dentists, and a growing—
but unknown—number of black midwives. The VNSP, Health Department,
and Phipps hired black nurses to care for black patients in their homes, and
white staff were expected to care for both white and black patients. Oversight at these three organizations, however, was clearly deemed a responsibility of the white community.[59]

HOME CARE IN PHILADELPHIA

The VNSP always cared for black patients, as indicated in the records by consistent notations of "colored" on lists of "nationalities" served. As early as 1896, 14 percent of the society's 1,255 patients were black. With massive immigration from Europe, the number of nationalities listed increased from fifteen at the turn of the century to thirty a decade later. Annual reports suggest that the chief preoccupation of the VNSP's board of lady managers was the health of the city's newest arrivals from Europe rather than that of the city's black population. In fact, other than their inclusion in annual statistics, black patients (much less race relations) received no special attention in VNSP records.[60]

By the early 1920s, Katharine Tucker, then superintendent of the VNSP, reported a "tremendous increase in the Negro population, with the attendant increase in health problems for the city as a whole." Typical of white expressions of concern about the menace of "Negro migration" was Tucker's warning the VNSP's white board of lady managers that the new "dangerous" immigrants no longer came from Europe—now they came from the South. As with earlier "dangerous immigrants," the board was reminded that "surely there is no sounder and surer way of helping these and other strangers within our gates to assimilate our standards and customs than through a health teacher in the home who shows through demonstration, as well as by precept, how health can be made possible."[61]

Not surprisingly, the VNSP's work with the black community did increase. For example, the 5,788 visits to the homes of black patients in 1922 represented an increase of 1,700 from the previous year. In 1927, however, both the Colored Women's Protective Association and Forrester Washington complained that "work among colored people is being neglected." The VNSP responded with statistics: 24 percent of its work was with black patients (7,048), costing the organization $60,000 per year. Tucker further noted that, within the past ten years, with the "colored" population increasing 100 percent, the caseload increased proportionally from 13 percent of the overall caseload in 1917 to 25 percent in 1926. Tucker's response reflected a national standard that suggested her agency's caseload should reflect the community served. Using this formula, at least 11 percent of the VNSP's caseload should be, and was, black. The VNSP confidently believed that it was serving the black community adequately, but the black community no doubt considered these services rendered "always with reservations."[62]

Katharine Tucker, characterized by Ethel Johns as friendly and "sympathetic to the colored group," supervised a staff of 113 nurses, of whom three were black. All were paid the same salary. While these black nurses were described as "reliable and industrious," Tucker thought their record-keeping work was poor and their nursing technique lacked finish. They were well received "by their own people who seem[ed] to take 'race pride' in the fact of their employment." Tucker's greatest objection to the employment of black nurses was the "geographic problem," that is, the wasted time in getting from one black district to another. Finally, she saw little hope of their employment as supervisors. "Naturally, they would not be permitted to direct white nurses, and if they were placed exclusively in charge of the colored staff, the question of 'segregation,' always so alive in Philadelphia, would at once be raised."[63]

In her report for the Rockefeller Foundation, Johns also described the work of black nurses employed by municipal health departments in Chicago, New York, and Philadelphia, where employment was not a "matter of preference, but because there [was] no way of excluding them under civil service regulations providing they [could] present the necessary qualifications and pass the required exam." The supervisors in all three cities "frankly" told Johns that they preferred not to employ black nurses, since they complicated the service and created "social friction." From the supervisors' perspective, white nurses did a better job and were "accepted by both races without question."[64]

Although Tucker's so-called sympathetic views were surely oppressive to black nurses in her employ, the even more palpable disrespect at Philadelphia's health department must have been a daily trial for black nurses. With a staff of thirty-eight white and four black nurses, the chief nurse of the health department, Mrs. A. H. Culbertson, said black nurses were "less than desirable in every way," except for their ability "to make contacts with their own people." They were unable to teach patients, showed poor judgment, were not punctual, and were of doubtful reliability. "They trade[d] on their superficial knowledge of theory to get them by—and upon their political influence to keep them employed." Finally, they complicated her administrative duties, because they only worked among their own people and could never be sent to white homes.[65]

"FOR THE HEALTH OF A RACE"

Not all evaluations of Philadelphia's black nurses were so negative.[66] At the Henry Phipps Institute, the first endowed center for the study, treatment, and extermination of tuberculosis in the United States, black nurses and doctors were credited with a dramatic increase in the number of black tuberculosis patients seeking care. This "novel" strategy made Phipps more "attractive to the patients." According to Henry Landis, "The Negroes no longer fear discrimination as they once did, but show entire confidence."[67] Prior to the employment of black nurses in 1914, the average caseload of black patients at Phipps was fifty. By 1921, the caseload of black patients had risen to 521, representing 1,743 visits to the clinic. By 1934, black patients seeking care at Phipps exceeded the number of white patients, with 3,263 black patients making 14,345 visits to the clinic and the staff of ten black nurses visiting 5,145 black homes. The Phipps experiment was so successful that it was replicated in several of the city's other tuberculosis clinics.[68]

A comparison of Phipps's use of black nurses for the black community in Philadelphia with Anna DeCosta Banks's work in Charleston suggests that, in both places, nursing skill and respectful teaching produced "unqualified success." The first nurse hired by Phipps was, like Banks, exceptional. Elizabeth Tyler graduated in the first class at Freedmen's Hospital School of Nursing in Washington, D.C. After graduation, she worked as a private-duty nurse in Northampton, Massachusetts, caring primarily for students attending Smith College. Hearing of employment opportunities in Alabama, she went South. She worked first at A&M College in Normal, Alabama, as the resident nurse and also taught physiology and hygiene. Moving to New York City to further her education through a postgraduate course at Lincoln School for Nurses, she also accepted a position at Henry Street Settlement in 1906.

As the first black nurse at Henry Street, Tyler was expected to find black patients in need of care. Using all sorts of creative strategies to gain community trust and confidence, she quickly made her way into black homes. After only three months, her caseload was so large that a second black nurse, Edith Carter, was hired. In December 1906, Tyler and Carter established the Stillman House Branch of Henry Street for Colored People in New York's San Juan Hill neighborhood. What Tyler learned in New York's neighborhoods about racial and ethnic conflict, distrust of the medical establishment, rampant disease, and reliance on traditional cures was great preparation for the challenges awaiting her in Philadelphia. Co-worker Lavinia Dock described Tyler and Carter as excellent nurses—especially alive to social movements and preventive work. By 1925, Henry Street employed twenty-five black nurses.[69]

Arriving in Philadelphia in February 1914, Tyler was hired by Henry Landis, director of the Clinical and Sociology Divisions at Phipps. Her job was to go into the community, find blacks who might have tuberculosis, and convince them to come to Phipps for diagnosis and care. While Landis's aim was clear, he had no idea how to proceed. His version of the story was that he and Tyler "fenced" for thirty minutes—"she trying to find out what I wanted of her, I wracking my brain to tell her." Tyler agreed with Landis that how one nurse would "tackle" eighty thousand people was a serious question—but she knew the answer.[70]

Described as a convincing talker, Tyler set out to accomplish what the white staff at Phipps had been powerless to achieve. She began in the neighborhood surrounding Phipps, known as the Black Belt of Philadelphia. She saw the household as the unit of her tuberculosis work, and her methods included friendly visiting, being present when her patients came to the clinic, and giving talks in neighborhood churches. Shortly, it was reported that for every patient Tyler sent in, five or six came on the advice of a friend. Black patients increased so quickly that by the end of her second year two more nurses were hired, their positions funded by the Whittier Center and the Philadelphia Committee of the Pennsylvania Society for the Study of Tuberculosis. A black physician, Henry Minton, was added to the staff in 1915. His salary was paid by the Pennsylvania State Department of Health.[71]

The Phipps story reveals the complexities of anti-tuberculosis work done for blacks by blacks. Landis and the Phipps Institute were successful because they understood the need to hire black visiting nurses who could gain the confidence of black patients. Nevertheless, care remained decidedly separate, with a segregated dispensary system where black doctors and nurses worked with black patients. Landis claimed that he entertained the possibility of turning over responsibility for this work to the black medical staff, but concluded that they still required "constant association with and supervision by sympathetic whites." Despite their ability to succeed where the white staff had failed, blacks were deemed not quite capable of self-direction. Reflecting the contemporary strategy of white-financed racial self-help and uplift, Landis chose to accommodate both whites and blacks. At the Phipps Institute, black physicians and nurses were respected professionals, black patients received care, and whites' anxieties about the "Negro problem" (spreading tuberculosis) were reduced.[72]

The "Landis model" was not accepted by all Philadelphians. Especially critical was black physician Nathan Mossell, founder of Frederick Douglass

Hospital; he believed care should be given without regard to race: "Where we make this problem a race problem, we defeat our ends." But Henry Minton, who practiced at both Phipps and Mercy Hospitals, disagreed. Minton believed that black physicians and nurses "know the innermost thoughts, ambitions, and peculiarities of their own people, and receive from them that response and confidence which would not be possible in a program from which they were omitted."[73] Although some debated integrationism and separatism as strategies for advancement, black health care providers and patients still found it difficult to move into positions of full participation in Philadelphia's predominantly white medical institutions.[74] A 1946 survey of hospitals in Philadelphia conducted by the local Association of Colored Graduate Nurses found a growing number of respondents hiring black nurses. However, "remarks" by those uninterested in hiring black nurses confirmed that, in Philadelphia, racism was ever present.[75]

North and South: "A Race's Problem"

In the patient's home, the intimacy, isolation, and nature of the exchange between patient and caregiver created a curiously unique situation. Success in providing care in the home required a sense of propriety and respect for local customs. Rules for caregiving (usually implicit) were based on gender and power, with physicians, supervisors, and lady managers in charge. Nevertheless, those in authority were rarely at hand for the act of caregiving, and no documentation tells what actually happened in the home. We can assume, however, that differences of race, class, and gender often complicated the development of mutual trust between provider and recipient, so essential if all were to feel safe.

In the South, reluctance to hire black visiting nurses as appropriate caregivers was a function of both local custom and law. In the North, it "worked best" to hire a black nurse if a sufficiently large and concentrated black population supported her. The only reported Northern attempt to send black visiting nurses into white (Italian) homes proved disastrous and was quickly discontinued. Poor white patients "resented the attention of a colored nurse" and would not accept black nurses as caregivers. In northern cities, the proportion of black staff varied widely (2% to 15%), the differences being attributed to political climate, size of the black population, racial outlook, employment opportunities, and the *possibility* of freer participation in society. In the South, black nurses did care for white patients but, as the

Charleston story illustrates, this practice was not always acceptable to working-class whites. By the 1920s, 3 percent (365) of the black nurses in the United States were employed in public health. Nearly 60 percent of these worked for tax-supported health departments and the rest primarily for voluntary agencies. Of these nurses, 159 lived in the South and 139 in three northern cities: Philadelphia, Chicago, and New York. By 1931, there were 549 black public health nurses in the United States; their geographic and employer distribution remained essentially unchanged.[76]

Employment opportunities for black nurses were limited, of course; jobs rarely turned over and white employers claimed that the supply of black nurses generally exceeded demand. In both the North and the South, employment opportunities were enhanced whenever the black community provided funds expressly to employ black nurses. As might be expected, job security in these positions was contingent upon the black community's continuing financial support.[77]

In the North and the South, white families occasionally hired black private-duty nurses, who often performed "domestic chores" in addition to caring for the sick, worked longer hours for lower wages, and were said to know "their place" in households and within the medical hierarchy. The acceptability of black nurses entering affluent homes was based not on their authority as trained professionals but on their place as servant employees. In most northern cities, private-duty nurse registries claimed little demand for black nurses. As one registry reported, "Negro nurses are never busy enough; just barely get along, and when the season is dull they are on the point of starvation."[78]

In the North, black visiting nurses were usually paid the same wages as white nurses. Southern traditions of racial discrimination dictated that black nurses be paid less than their white co-workers. Pay inequities were allegedly based on the inferior training of black nurses and a belief that blacks could "live more cheaply." In the South, employers considered it "manifestly unfair to pay the same salary to a woman whose living expenses could not amount to much more than half those of other women in the same work."[79]

Assorted strategies were employed to meet the health needs of the black community, and the tenor of these varied from place to place and from organization to organization. The MLI statistician Louis Dublin believed northern migration had negatively affected black health prospects and that black "energies" would have been best served if blacks had stayed down south on their farms. Although some southern health officers looked forward to the "ultimate extinction of the colored race" from tuberculosis, syphilis, and maternal and infant mortality, few nurses shared any enthusiasm for this approach to "race adjustment." White nurses' descriptions of their black patients convey a message of disdain and inferiority intermixed with an understanding of struggles to survive in crowded and unsanitary conditions. Nurses recognized that "germs have no color line" and understood white self-interest. As the popular saying went, "A chain is as strong as its weakest link, and since diseases are contagious and infectious, the safety of the many depends on the health of the few." If black health also meant white health, someone had to go into the homes of black patients to provide care and prevent the spread of disease. It is little wonder that black nurses were seen as an indispensable solution to "a race's problem," especially when bedside care in black homes required physical contact or exposure to dangerous diseases. In both the South and the North, black nurses were increasingly acknowledged as "especially adapted to bedside care" and respected for their "ability to establish contact with their own race."[80]

In both North and South, it was mutually agreed that black nurses had their proper place. With regard to home care, distinctions between the races clustered around a very short list of variables: bedside care, teaching ability, capacity to motivate, supervisory competence, and quality of training. No matter how well trained they were or what their experience or expertise, black nurses felt that in the "white mind" they were intruders into the profession. "Brow-beaten . . . looked upon in an attitude little better than one would look upon a snake or an insect," black nurses hoped eventually to be judged by their merits rather than the color of their skin. Given their status as inferior members of the nursing profession except when caring for black people like themselves, it is amazing that black women ever "answered the call" to nurse.[81]

Compliments by whites about black nurses were at best condescending and almost always nuanced with perceptions of black racial characteristics. Black nurses were described as "by nature wonderful nurses, kindly, cheerful, dexterous with their hands, of musical voice and comforting ways. They are intensely loyal with understanding born of long suffering." The black nurse interpreted domestic complications that would baffle a white nurse "who lacked her intuitive understanding of racial characteristics." The ability of black nurses to communicate, motivate, and gain confidence—"get a proposition over" to black patients—was most impressive. Unlike white speakers, who delivered messages politely received by black audiences, the eloquent program of health education delivered with "dramatic instinct" by black nurses left audiences shouting fervent amens to "summons to repentance and reformation."[82]

The training of black nurses was considered inadequate, especially for public health work. Many whites called for race advancement through replacement of the "inferior Negro" with a better educated nurse. To succeed, black public health nurses needed to be trained "in the knowledge of public work, but not overtrained until they lose the characteristics of their race, for the ignorant Negro is nearly as difficult to reach by individuals of this type of his own race as he would be by nurses of the white race." Clearly, it was acceptable to train black nurses in public health as long as they did not become too white or too high class or forget their place.[83]

Whites believed that black nurses lacked intelligence, experience, and education and therefore required more supervision, especially with "mixed" staff. Many thought black nurses were poor at clerical work, lacking in initiative or executive or organizational ability, and incapable of leadership. Against this backdrop of racial prejudice and the "weight of tradition," it was widely understood that black nurses could not become supervisors. In 1930, of the 287 black nurses employed in northern public health organizations, only six held supervisory positions: one in Chicago, four in New York City, and one in Philadelphia.[84]

Most opportunities for advancement in nursing required special training in public health, and most postgraduate courses accepted black nurses with certain reservations, the extent determined by the local situation. In other words, black students could be sent only to acceptable places—meaning black homes. Scholarships for black nurses were in short supply. Even the Julius Rosenwald Fund—which between 1917 and 1940 donated over $1.3 million to support professional education, public health, outpatient services, and hospital care for the black community—invested in black nurses only when intellectual caliber was deemed high and a future was definitely guaranteed, that is, for those returning to assured positions. In the absence of opportunities for promotion, it is not surprising that a 1930s survey found a general disinclination on the part of black nurses toward advanced preparation. A decade later, of the estimated 25 percent of black nurses (623) working in public health, 52 percent had some postgraduate training and only 20 percent held positions above a staff level.[85]

A telling reconstruction of white nurses' thoughts on race betterment is seen in Mary Beard's recollections of a 1926 trip to the South made on behalf of the Rockefeller Foundation. In her "private" field notes, Beard tells of an exhilarating morning spent with "Nurse Volo (colored)." It began with several perilous slides on rain-slicked dirt roads in the mountains, near catastrophes crossing narrow plank bridges, close encounters with "mules, Negroes, and carts," and collisions with three cars and one tree. Although Beard found her companion's driving "somewhat casual," she was very impressed with "the value of Nurse Volo's visits." Volo "displayed tact, force, intelligence, and exhibited that real devotion which was returned with enthusiasm." She was able to influence a contract doctor to do the right thing, and her "plain intelligent instruction" on a prenatal visit to a mother with a delicate child, for a baby with ophthalmia neonatorum, and to a young couple with gonorrhea "were remarkable because so effective."[86]

In Beard's published account, Nurse Volo became Nurse Emma, the patients backward, the ride exciting, and the physician nowhere to be seen. Setting the stage for the standard educational and oversight strategies for race betterment, Beard's revised narrative attributed Nurse Emma's unusual aptitude for teaching and her adaptability to difficult home situations to an understanding acquired through personal success with similar experiences. Adding the requisite examples of Nurse Emma's ignorance, Beard concluded that Nurse Emma's instinctive good works "would have profited by special preparation in public health nursing and adequate public health nurse supervision." With this great and expanding "field for Negro nurses among their own people," Beard wondered, "why do not more of the best women of the race prepare themselves for public health nursing among their own people?" Had Nurse Volo been given the opportunity to read Mary Beard's publication, "Nurse Emma," there is little doubt she could have offered some "lived" enlightenment to Nurse Mary.[87]

In Their Own Words

The voices of black nurses were rarely heard by their white nursing colleagues. Their writings occasionally appeared in the nursing journals and accounts of their work were usually reported by others. However, as suggested by Gloria Smith, a contemporary black nursing leader, "Black nurses were accountable to black people in a special way." And Louise Nelson, former nurse-administrator at Freedmen's Hospital, wrote in 1928, "Black nurses wanted to do their part!"[88]

Thanks to the recent work of several historians, we now know that, in

striking contrast to white discrimination, demeaning treatment, and negative stereotypes, the black community held black nurses in high esteem.
Confronted by disease, poverty, and racism, black nurses wished to do their
part for those in distress or those who wanted and needed a helping hand.
Like their white colleagues, black nurses claimed a heritage of enlightenment, love, and service. As historian Marie Mosley reminds us, they found
"creative ways to transcend . . . barriers and overcome circumstances that restricted their practice and destroyed their patients' lives."[89]

The few published accounts by black nurses illuminate several aspects of
race relations particular to home care. An outstanding example of what can
be learned from black versus white versions of these stories is Jessie Sleet's
description of her turn-of-the-century work in New York City. The life of
Canadian-born Jessie C. Sleet (married name, J. S. Scales) is reminiscent of
the experiences of Anna DeCosta Banks and Elizabeth Tyler. Sleet is said to
be the first black visiting nurse in the United States. Trained at Chicago's
Provident Hospital School of Nursing, she graduated in the fourth class in
1895. She moved to New York City intent upon becoming a visiting nurse.
Sleet described her reception at Mission Societies and Bureau of Charities,
an organization known to have a "friendly attitude toward the colored race,"
as pleasant and interested. She nevertheless remained unemployed for several months.

Eventually Sleet was referred to Edward Devine, social reformer and general secretary of the Charity Organization Society. He was apparently deeply
impressed by Sleet and most distressed to realize that, in his organization,
"no colored were employed . . . not even as cleaners." Discovering he was
alone in his enthusiasm for Sleet, Devine identified a benefactor, Herbert
Parsons, who was willing to fund a two-month experiment. On 3 October
1900, Sleet cautiously accepted this provisional job as a visiting nurse for the
black community. After a successful year, Sleet obtained a permanent job
and her name was recorded in the society's books. She continued in this position for several years until she married, and was then asked to train her
successor, Cecile Batey Anderson, a graduate of the Lincoln School for
Nurses.[90]

A report of Sleet's work appeared in the first issue of the *American Journal of Nursing* in a section titled "Progressive Movements." The article, "A
Successful Experiment," was written by Sleet but submitted without her
knowledge by an admirer (perhaps Lavinia Dock) of her altruism and intelligence in social reform work. After her calls to destitute and sick patients,

Sleet concluded that "this house-to-house visiting, these face-to-face practical talks, which I have been having with the people, must be bringing good results. They have welcomed me to their homes, saying, 'We don't know you, but we belong to the same race.' They have listened to me with attention and respect, and if advice which I gave was not always accepted, in no case was it rudely rejected."[91]

Sleet acknowledged that conditions among her race were "peculiar" and therefore "peculiar methods must be taken." Patients were described as southern, poor, living in crowded conditions, and having inadequate diets. Having experienced both economic and racial exclusion from health care, most believed that professional care was required only when they were confined to bed and unable to work. Initial responses to illness usually incorporated a variety of self-help strategies: nontraditional and patent medicines, faith healing, and ritual cures. White medical professionals were typically distrusted and feared. Stories of death-producing "black bottles," tuberculosis patients forcefully removed from their homes to hospitals, and the authoritative and punitive styles of health care workers contributed to distrust and poor communications.[92]

Sleet concluded that "the insurance companies were largely responsible for much of the secrecy regarding the true condition of the patient. Some of them refused to insure colored persons on the grounds that they were more susceptible to pulmonary diseases than whites, while for the same reason, others required higher premiums and still deducted a portion from the policy when the insured died of tuberculosis."[93] Like millions of working-class Americans, many blacks were eager to buy these policies to ensure that family members could pay burial expenses and have a few dollars to begin again. For years, black mortality rates made such insurance policies a poor business investment. Even when insurance companies cautiously wrote these policies, blacks tended to get less insurance and at a higher price than whites. The black community understood what was meant by insurance concepts of "Negro risk," "careful selection," "good moral risk," "class distinctions among Negroes," and "extra mortality," and carefully avoided disclosing too much information to the wrong person.[94] (For a more extensive discussion of insurance, see Chapter 7.)

The description of Sleet's care of patients was indistinguishable from that of her white colleagues, but her approach to the community was far more dynamic. Beyond knocking on patients' doors, she also visited societies, churches, physicians, ministers, and even undertakers to explain her work and obtain trust and cooperation. Hers was a strategy of respect for and understanding of how her community worked. Unlike many of her white coworkers' visits to black homes, Sleet's visits were welcomed and never considered an intrusion into the family's privacy.[95]

Thirty years later, in 1934, Midian Bousfield, a black physician, Director of Negro Health for the Julius Rosenwald Fund, and vigorous spokesman for the black community, would echo Sleet's message. As the first black person to address the American Public Health Association in its sixty-year history, he gave a clear message: neglect of the black community's health needs and gratuitous racism were unacceptable. In his talk "Reaching the Negro Community," he told his white audience that they needed to learn the right way to work with the black community. In every community certain amenities and certain standards plagued the uninitiated. First, whites needed "a little prep in the psychology of the Negro community." The community was not unorganized; it had leaders and pseudo-leaders, a press, and organizations. Whites also needed to read the black press, talk with black public health workers ("including nurses"), work with black businesses and social service agencies, and visit barbers, movie houses, clubs, and churches. Arguing that the interracial approach was essential and useful, he warned whites to watch their language. Common white errors that had "gained fame" in the black community should be avoided: make no reference to the race question; leave out previous experiences with black people; forgo any expression regarding a lack of prejudice; omit the "darky" story in dialect; talk as you would to any neighbor; and, if possible, present statistics especially related to blacks. As Bousfield reminded his audience of white public health workers, by participating in interracial experiments for social change, they had an opportunity for "brilliant achievement," fulfillment of moral responsibility, and fame. Bousfield concluded, "The best way to begin is to begin."[96]

In both North and South, the peculiar realities of home care required that the interchanges between white and black patients, families, and caregivers be dictated by the "stiff-sided box" of race, class, and gender. While white insensitivity and racism improved over time, the peculiar variations from subtle prejudice to outright discrimination persisted throughout the North and the South to the present day. As we have been reminded by several recent studies, race, class, and gender are powerful predictors of access to and outcomes of health care.[97]

CHAPTER 5

Lillian Wald and the Invention
of Public Health Nursing

O ver one hundred years ago, Lillian Wald coined the term *public health nurse,* hoping to prescribe a new role for nurses who visited the homes of the sick poor. Her vision of public health nursing was service to the community by working to improve standards of living, "personal and civil." Like her visiting nurse colleagues, Wald understood that home nursing must be undertaken seriously and adequately on terms most considerate of the patient's dignity. Her persistent articulation in public forums that sickness should be considered in its social and economic context led to innovative and pragmatic remedies for seemingly overwhelming problems. According to Wald, "The call to the nurse is not only for bedside care of the sick, but to help in seeking out the deep-lying basic causes of illness and misery, that in the future there may be less sickness to nurse and cure." As the definitive public health nurse, Wald was instrumental in securing reforms in health, industry, education, recreation, and housing, and she originated the ideas leading to establishment of the Children's Bureau, school nursing, insurance payments for home-based nursing care, and rural nursing (through creation of the Red Cross Town and Country Nursing Service).[1]

What Wald called "our enterprise [of] public health nursing" was not an isolated undertaking, nor was she a lone American heroine. Her paradigm for nursing practice was based on knowledge gained during two decades of experience in visiting nursing and owed much to the Progressive reform and public health movements of the turn of the century. Historically, Wald is characterized as a visionary of legendary accomplishment. Although Wald embodied the professionalization of visiting nursing, due credit must be accorded to the thousands of nurses who legitimated the practice of nursing in the community.[2]

Journey to the Lower East Side of New York City

Lillian Wald was born on 10 March 1867, to parents described as descendants of many generations of rabbis, merchants, and professional men. Her father's success in the optical business allowed Wald and her siblings to enjoy a happy and affluent life in the German Jewish community of Rochester, New York. Although her parents supported charitable causes, they were not champions of social reform, and Wald grew up with little exposure to the lives of the immigrant poor.[3]

At the age of twenty-two, Wald chose to enter the nursing profession. Finding her life of society, study, and housekeeping unsatisfying, she wished to pursue serious work. In her application to nursing school, Wald professed a natural aptitude for nursing, which had for years appeared to her "womanly, congenial work, work that I love and which I think I could do well." After she graduated from the New York Hospital School of Nursing in 1891, her first job, at the New York Juvenile Asylum, left her dismayed, forever embittered toward institutional care for children, and discouraged about nurses' ability to produce institutional change. Uninterested in working as a private-duty nurse in the homes of the wealthy, she decided to go to medical school.[4]

After a disheartening start, Wald's story takes off in the winter of 1893. As a newly enrolled medical student, she learned that a Sabbath school for immigrants needed a course in home nursing. Although Wald was utterly ignorant of the realities of life for immigrants on the lower East Side, she nonetheless agreed to establish and teach the class. On 6 March 1893, she made the journey to the home of her student, Mrs. Lipisky. "Within a half hour" the course of nursing history changed, as Wald was guided by Mrs. Lipisky's young daughter through crowded "evil-smelling" streets, past open courtyard "closets," up slimy tenement steps, and finally into the sickroom. Wald later described this experience:

> All the maladjustments of our social and economic relations seemed epitomized in this brief journey and what was found at the end of it. The family to which the child led me was neither criminal nor vicious. Although the husband was a cripple, one of those who stand on street corners exhibiting deformities to enlist compassion, and masking the begging of alms by a pretense at selling; although the family of seven shared their rooms with boarders . . .

and although the sick woman lay on a wretched, unclean bed, soiled with a hemorrhage two days old, they were not degraded human beings, judged by any
measure of moral values. In fact, it was very plain that they were sensitive to
their condition, and when, at the end of my ministrations, they kissed my
hands . . . it would have been some solace if by any conviction of the moral un-
worthiness of the family I could have defended myself as a part of a society
which permitted such conditions to exist. Indeed, my subsequent acquain-
tance with them revealed the fact that, miserable as their state was, they were
not without ideals for the family life, and for society, of which they were so
unloved and unlovely a part.[5]

This was for Wald a baptism of fire. Naively convinced that such conditions were possible only because people did not know they existed, she committed herself to learn and to tell. Wald and Mary Brewster, her comrade
from nursing school, devised a plan to live in the neighborhood as nurses.
To support their plan, they sought financial backing from Mrs. Solomon
Loeb, who had supported the Sabbath school classes. Loeb found Wald to be
an extraordinary young woman, either a "great genius or mad." Preferring
to think her a genius, Loeb and her son-in-law, Jacob Schiff, agreed to underwrite Wald and Brewster's neighborhood nursing plan for six months.
Free from every form of control, Wald and Brewster's mission was to explore
and discover, "to do what they could; to see what they could see; and to publicize all that was wrong and remediable."[6] Having found an acceptable way
to practice nursing, Wald never felt the need to return to medical school.

Jacob Schiff secured the endorsement of the New York City board of
health, which provided badges identifying the nurses as visiting nurses under the auspices of the board of health. Schiff made arrangements with the
medical staff of the United Hebrew Charities for professional consultation
and assistance. Two years later he would provide Wald with the house on
Henry Street, thus enabling her to develop her many ideas.[7]

Schiff's financial backing coupled with Wald's vision of visiting nursing
allowed for a more radical agenda of social reform. Wald would maintain
Loeb and Schiff's interest in the years ahead by making them aware of the
dire circumstances in which her patients lived. Schiff and his wife, Theresa,
often dined at Henry Street, seeing en route those struggling immigrants
whose lives could be elevated by their generosity.

Instead of the traditional printed annual reports used by most visiting
nurse associations, Wald made real the names and faces of the needy. Her
early reports to Schiff and Loeb, beginning when she and Brewster lived in

their first tenement home on Jefferson Street, were simple, quickly written letters describing tragic conditions. For example, Wald wrote to Schiff on 2 July 1893 about her visits to twins with measles, a baby with ophthalmia, another with vermin bites, a man deformed by rheumatic disease, and a woman who supported her family by nursing an infant whose mother had died. Her report to Schiff included the following story:

> No. 7 Hester Street proved two tenements, climbing the stairs in search of [the mother of the child with ophthalmia] found terrible filth everywhere, sinks filled with slop, floors reeking. I went into every room in the front and rear tenement, set the dwellers to sweeping, cleaning and burning refuse. In some rooms swill thrown on the floor, vessels standing in the rooms unemptied from the nights use. I saw the housekeeper, who promises cooperation in keeping the place cleaner and I impressed him that I would repeat the rounds the next day and frequently after. The house had many nursing infants, seven of whom had summer bowel trouble—one child, 3 weeks old with measles and badly excoriated buttock. I bathed this child, teaching the mother how to do it properly, to care for the mouth and left swabs and the essentials for its proper care. Next room, Sarah K. showed her babe 15 months of age with bowel complaints, but child was clean and being properly cared for, so I did nothing but give an ice ticket.[8]

Over the coming years, Wald's growing enterprise was financed by a small group of generous benefactors whose continued support required constant maintenance. Despite regular disagreements and political strains, their support allowed Wald to experiment with a wide range of innovative programs. This contrasted with the funding of the majority of visiting nurse associations, which depended on the good will of hundreds of small donors for financial support. In most associations, embarking on risky experiments required the endorsement of the board of lady managers, which made innovative programs or modifications in the nurse's role far more difficult.[9]

An Untrammeled and Spontaneous Enterprise

Unlike the documents left by some of her peers, Wald's writings provide little insight into the thinking that produced simple yet functional solutions to a multitude of complex problems. Her biographer, Robert Duffus, maintains that she moved to the lower East Side with "no theories about economics, sociology or politics, little knowledge as to how people outside her own social group lived, no panacea to try out, no sweeping vision of the future . . . but she did have an imagination which enabled her—more than that, compelled her—to put herself in other people's places."[10]

Despite claims of the "untrammeled and spontaneous character of [her] enterprise," Wald's attendance at the International Conference of Charities, Corrections, and Philanthropy at the 1893 Chicago World's Fair may also have shaped her thinking. Here, Wald met influential women reformers from England and from elsewhere in the United States. Papers written by Florence Nightingale and Mrs. Dacre Craven, whose ideas on nursing Wald had studied during her training at New York Hospital, were especially fitting to her plans. Wald also met Lavinia Dock, who later visited Wald and Brewster on Jefferson Street and in 1896 joined "the family" at the Nurses' Settlement on Henry Street. Dock became one of Wald's closest friends, her confidant, and editor of her writings.[11]

Inspired by the International Conference, Wald and Brewster moved to Jefferson Street on the lower East Side in July 1893, thus joining the growing ranks of "new women." Described as a revolutionary demographic and political phenomenon, the new woman was typically single, educated, and economically independent, a champion of professional visibility for women and an advocate of social and economic reform. Like these contemporaries, Wald defied the rules of social propriety while retaining a place within their genteel world, at the same time seeking and achieving considerable political power.[12]

While later admitting to "youthful heroics," Wald claimed that "all that we ever did we did the first day." With experience, Wald and Brewster's ideas, techniques, and administrative methods crystallized, but the "test" that provided their vision was imagining how "their loved ones would react were they in the place of the patients whom we hoped to serve." Gradually, it became clear that a system for nursing the sick in their homes could not be firmly established unless it introduced a new etiquette of how visiting nurses, physicians, and patients interacted with each other in the community. Wald's redesigned professional conventions proposed that (1) access to nursing care at home should be determined by the patient's need unencumbered by the red tape or formality of a hospital's or doctor's rules; (2) nursing of the sick in their homes should be undertaken seriously and adequately, and instruction of patients in the ways of healthful living should be an incidental, not a primary, consideration; (3) care provided to the poor should be indistinguishable from that received by those who could afford to purchase care of their own choosing; and (4) services should be created on terms most considerate of patients' dignity and independence.[13]

Wald claimed that this new system of visiting nursing immediately changed the nurse-patient-doctor relationship. Referrals came most often from patients' families, although some were from charitable organizations, clergy, and physicians. A nursing visit usually preceded a call to the doctor, with the nurse deciding whether the patient needed medical assistance at a dispensary, "uptown specialists," or hospital care. No longer was the nurse a visitor whose presence proclaimed to neighbors the family's poverty, acceptance of charity, or even worse, inability to care for sick family members. Now the nurse's appearance announced a commitment to do everything possible for the sufferer without loss of self-respect. Wald hoped that a nursing mission fashioned by the concerns of a few urban "lady bountifuls"— who believed that an occasional lesson about germs, boiled water, and sterilized milk could transform the poor into "little bowers of thrift and neatness"—would be supplanted by a mission that better appreciated the "social facts" of the times. The old-fashioned nurse, who in the words of socialist Lavinia Dock was simply a bridge between the rich and the poor, a "sop thrown from capital to labor," was to be replaced by Wald's new vision.[14]

It was in recognition of the relationship between illness and poverty, placing disease beyond individual control or escape, that Wald coined the term *public health nurse:* cure by nurses, Wald believed, mandated treatment of social and economic problems, not just illness. The nurse, through her particular introduction to the patient and her democratic and neighborly involvement with the whole neighborhood, was the starting point for wider service. Thus, the nurse and her agency never worked alone; they were linked with all agencies and groups working for social betterment.[15]

Wald's Experiment in Public Health Nursing

Lillian Wald and Mary Brewster arrived on the lower East Side during one of the nineteenth century's worst depressions, and there they confronted disease, vermin, poverty, "tenement intelligence," and the filth of crowded tenement homes. Recognizing that the sickness encountered in families was part of a larger set of social problems, Wald immediately began to mobilize an impressive if disjointed array of services, from private relief agencies to the medical establishment itself. Creating cooperative relationships with organizations as varied as hospitals and newspapers allowed her to provide patients with ice, sterilized milk, medicines, meals, and referrals to many of the city's hospitals, dispensaries, and, most important, jobs. Hospitals, dispensaries, and relief agencies, hearing of the nurses' work, became "believers" and sent patients for care and follow-up.[16]

Wald and Brewster quickly uncovered one of home care's major dilemmas—the impossibility for families of adding caregiving to the other daily demands on their time. Like all their visiting nurse colleagues, they found it essential to send a housekeeper to scrub, cook, provide child care, and clean up patients' homes so the breadwinner could continue working. As word of the nurses' work spread, donations of food, sputum cups, medicines, car fare, clothes, bedding, sick room utensils, surgical dressings, cases of soap, and some money began arriving regularly. Anxious to avoid "pauperizing" their patients, the nurses did not intervene as long as the family scraped along. Only when absolutely necessary did they arrange for loans or for temporary payment of patients' rent.

Wald herself washed dirty patients, claiming "the use of scrubbing brush and bath has been frequent—if we can't assert it to be permanent," and rewarded any real household or personal "transformations" with new clothes or linens. Under the authority of the board of health, she began a neighborhood-wide campaign to clean roofs, disinfect "vaults," and clean tenement hallways. "Greater nuisances," such as offensive stables, were referred to the board of health as part of the nurses' daily reports.[17]

By November 1893, Wald was reporting more cases of acute illness and need for operations, and by January the tragic condition of their neighbors showed no sign of relief. As the depression continued into the winter, she learned firsthand the face of famine, and she and Brewster spent a great deal of time "bringing relief or reporting for relief the better class of idle working men, who, too proud to ask for charity, were virtually starving." As sad as "this winter's story" was for Wald, she remained thankful for friends and benefactors who made some relief and care possible. Seeing every day "enough sorrow and poverty and illness to fill a world with sadness," Wald said she could not have lived through the sights they saw if she had not been able to help.[18]

The depression emphasized to these nurse-witnesses the significance of employment, which brought the most desirable form of relief to the households they visited. Through Schiff and Josephine Shaw Lowell, founder of the New York Charity Organization Society, Wald was able to dispense jobs to some families under her care. She hired neighborhood women as housekeepers to clean patients' rooms and referred neighbors for jobs to a variety of other employers. But what they provided was not even faintly adequate.[19]

Two years of practicing among people without economic resources quickly convinced Wald that disease most often resulted from causes beyond individual control or escape, that treatment needed to be prescribed in an "all round way" with consideration for both the social and the medical aspects of the case, and that families should realize that outsiders could not carry the entire responsibility but rather provide guidance and some alleviation of problems. Wald characterized this as serving from the patients' point of view. She believed that, to succeed, the practice of public health nursing required democratic and neighborly relationships with those being served.[20]

Eager for their neighbors to have knowledge of a life beyond crowded tenements and factories, Wald regularly arranged picnics, social excursions to the country, and tickets for concerts, and ensured that those participating had a "respectable appearance" for such events. She coaxed mothers to allow their children to go to school and when unsuccessful, occasionally sent truant officers to visit. Opening little bank accounts in neighborhood children's names, she deposited any extra money the children earned. Despite her critics, including political activist Emma Goldman, who dismissed these as unimportant activities contrived to "teach the poor to eat with a fork," Wald persisted. Over Wald and Brewster's first holiday season on Jefferson Street, the neighbors reciprocated the nurses' hospitality by coming to "sing to us and some brought . . . little cards, passed most apparently through more hands than one pair." By Passover of 1895, Wald reported that the neighbors still visited, choosing to take "tea with us as their celebration." She concluded, "We are not tired of them nor they, apparently, of us."[21]

Wald's vision resulted in nursing practice that went well beyond simply caring for families during illness, encompassing an agenda of reform in health, housing, industry, and education. Wald thought it important to learn whether patients' problems were isolated or common to many, because the "technique for finding out" often led logically to identifying an appropriate remedy. From the beginning, every incident that "seemed to have community bearing was noted and held in reserve" until it could be broadcast so that others with greater resources could influence and educate public opinion toward assuming mutual responsibility.[22]

The House on Henry Street

In 1895, wishing to create a larger and more formal organization, Wald moved out of her tenement home into a nearby house that would become the Henry Street Nurses' Settlement. At that year's meeting of the Interna-

FIGURE 10. Members of the Henry Street Settlement "Family," circa 1900.
Seated (left to right): Mary Magoun Brown, Lavinia Dock, Lillian Wald,
Yssabella Waters, and Henriette Van Cleft. Standing (left to right):
Jane Hitchcock, Sue Foote, and Jeanne Travis. In front are two
Henry Street children, Sammie Brodsky and Florrie Long.
*Visiting Nurse Service of New York, Center for the Study of the History
of Nursing, School of Nursing, University of Pennsylvania.*

tional Conference of Charities and Corrections, Wald enrolled six zealous
women of talent, personality, ability, and spirit to become the staff at the
Henry Street house. She hoped the settlement's first members would bring
distinctive training and connections from a variety of nursing schools yet
create an esprit de corps essential to cooperative work (Figure 10). The settlement nurses were given a fellowship but collectively shared living expenses.[23]

Wald and her new staff moved to 265 Henry Street in the spring of 1895.
Within a short time the house had eleven residents. The family consisted of
a powerful group of nurse pioneers: Lavinia Dock, one of the first to arrive,
who brought her commitment to feminism, women's suffrage, and union

organizing; nurse-educator Adelaide Nutting, who in 1910 would become chair of the Department of Nursing Education at Teachers College, Columbia University; Annie Goodrich, Director of Nursing for the settlement; Lina Rogers, the first school nurse; and Yssabella Waters, visiting nursing's first statistician. Members of the laity, as the nurses called them, included Helen McDowall, who used her wealth to create Henry Street's theater and arts programs, and social activist and lawyer Florence Kelley, who would become general secretary of the National Consumer League. By combining the "natural simplicity of the household's rich fellowship" and regular frank discussions of "questions of current importance" with a disdain for conclusions based on majority rule, this diverse group of opinionated women lived and worked together. Although household members changed over the years, 265 Henry Street was always home to a spirited group of women committed to health care, employment, immigration, education, housing, trade unions, and other issues. Wald would later describe the settlement as the most pliable tool for social change ever developed.[24]

Daily rounds for members of the family began with breakfast at 7:30 a.m. They read mail, reviewed the day's work and plans, and discussed knotty problems and difficult situations. Wald assigned new patients, mindful of the size of each nurse's caseload, but nurses were responsible for managing their patients and their time, free from any restrictive regulations. The nurses customarily returned to the house for lunch, and when afternoon visits were completed took up their own special work—teaching English, home nursing, or perhaps a club for teenage girls. A cooking class, the Good Times Club, quickly became a favorite with the neighbors, despite its cost of a nickel per week.[25]

The house was also used to host controversial events unwelcome in more conservative settings. For example, when the National Negro Conference met in 1909 to examine the status of black people, the opening reception was held at Henry Street Settlement—one of the few places willing to host an interracial gathering. Concerned that an interracial sit-down supper would cause bad publicity, the event's planners decided that Henry Street was so small that all two hundred guests "would have to stand for supper." The party was successful and the ultimate outcome of the conference was the founding of the NAACP.[26]

One room of the settlement served as the nurses' dispensary, where simple complaints and emergencies not requiring referral elsewhere were treated, and a bathroom was available for regular and emergency use by the neighbors. The

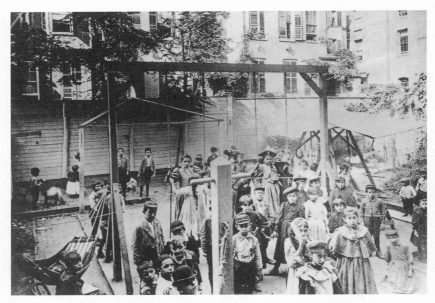

FIGURE 11. Playground, Henry Street Settlement, circa 1900.
*Visiting Nurse Service of New York, Center for the Study of the History
of Nursing, School of Nursing, University of Pennsylvania.*

settlement organized classes in home nursing, elementary first aid, household
hygiene, and child care for tenement mothers, girls, and boys. The yard was
converted into the "largest playground" on the lower East Side, with preference given to convalescent and crippled children (Figure 11). Wald installed
swings, slides, and sandboxes in the backyard. An arbored canopy along with
tables, chairs, and lemonade were provided for the mothers. One of the earliest urban playgrounds, it became known as Bunker Hill. To make the settlement a social center for the neighborhood, the house was comfortably furnished and supplied with books, pictures, music, and educational aids.[27]

During these early years, the settlement issued no public or formal reports
nor did it need to make appeals for money. Patients were encouraged to pay
what they could, but most care was provided free. Any fees collected from
patients went into an emergency fund for expenses incidental to the nursing service, such as car fares and supplies. The settlement also received some
income through contracts with several lower East Side lodges and benefit
societies to provide nursing care for their members.[28]

Collecting data, conducting experiments, and studying individual and community health quickly proved a successful technique. Wald's and the settlement's achievements, from sweatshop reform to experiments examining the feasibility of caring for patients with contagious diseases in crowded tenements, have been discussed extensively elsewhere.[29] Classic examples of their simple solutions for complex problems include insurance payments for home care (discussed in Chapter 7) and a less well-known innovation, the first aid room.

The First Aid Room

Henry Street Settlement's first aid rooms were developed based on a combination of factors: innovative analysis, circumstantial opportunity, and adaptive problem solving. Although most of the nurses' patients were sick at home, neighbors with minor problems found the nurses next door a quick source of care. Increasingly, neighbors stopped by for advice or care of fresh cuts and bumps, old wounds, eczema, burns, local infections, small accidents, conjunctivitis—"nursing cases as might be attended to by members of the families if the mothers had sufficient leisure or sufficient intelligence." By 1900, demand was so great that three first aid rooms were opened in response to neighborhood needs. It quickly became obvious that an enormous amount of such work was needed. From the nurses' perspective, this work was "hardly important enough to receive attention in the large, crowded dispensaries, where the patient [had] to wait for hours for treatment, yet really important to the general health." With nurses at hand, mothers could easily run in with their children, working people could come by after hours, and problems that typically went untreated received attention. Hoping to avoid sanctions by the medical profession, Wald rather cavalierly claimed the nurses were not running dispensaries in the strict sense, since no doctors were in attendance and no medicines were supplied. She contended that many of the cases were not important from a medical point of view and that patients needing the attention of a physician were, of course, always referred to the proper place.[30]

As the number of ambulatory visits grew, the settlement risked attracting the unwelcome attention of the increasingly disagreeable "uptown docs." The New York Medical Society's recent success in attaching a clause to the Nursing Registration Bill prohibiting nurses from practicing medicine gave the society a new opportunity to disrupt the settlement's neighborly activities. While initially the first aid rooms went unnoticed, by 1904 even Lavinia Dock was worried. Despite her usual bravado about nurses' independence, Dock wrote to Wald about doctor's concerns that nurses were carrying ointments and even giving pills outside the strict control of physicians.

Dock unequivocally claimed that the nurses were not engaged in such activities, nor did they want to practice medicine, but she nevertheless saw the need for having some standing orders—emergency treatment and medication orders endorsed by the local medical society—behind them. Maintaining that the nurses' work was for the benefit of patients, not doctors, the settlement had no intention of neglecting the growing demand for neighborly first aid, but they were also not interested in a potentially costly confrontation with the medical profession. By establishing standing orders, the nurses protected themselves from legal attack. Published statements, such as that a nurse should never prescribe for a patient, were widely circulated. The annual report now called these visits to the first aid rooms "office dressings." By 1911, all "proper dispensary cases" were eliminated from the first aid rooms and only patients sent by doctors or nurses were treated. Later publications reminded staff and assured the public that "the real Henry Street Settlement nurse will make the doctor feel that she is exerting every effort to have his treatment, not hers, intelligently followed," and that settlement nurses always worked under the direction of a physician. Protected by standing orders, policy, and the necessary public rhetoric, over the next fifteen years the settlement cared for between ten and twenty-three thousand patients in first aid rooms each year.[31]

Nursing and Health

By 1910, with a phenomenal increase in demand, the nursing service was outgrowing 265 Henry Street. Additional houses were opened throughout the city, and eventually the nurses began to live in flats of their own in the neighborhoods where they worked. In twelve posts across the city, a staff of fifty-four nurses ran a milk station, a convalescent home known as "The Rest," and three country homes with a capacity of seventy-five beds. During summer months, they also ran additional country programs for 120 children and young adults. Caring for 19,492 patients, the staff of fifty-five made 175,953 home visits and gave 18,934 first aid treatments (office dressings) in 1911. Nearly 40 percent of the settlement's calls now came from doctors and less than 20 percent from families and neighbors.[32]

Based on Wald's social philosophy, three kindergartens and classes in carpentry, sewing, art, diction, music, and dance were also instituted. Boys' and girls' clubs had a combined membership of twenty-five hundred. Staff and volunteers oversaw men's, young women's, and mothers' clubs, a dramatics group with its own theater, two large scholarship funds, and numerous informal activities. In December 1912, attendance at all programs reached twenty-eight thousand. As Wald noted in 1915, the work was demanding: "Enthusiasm, health, and uncommon good sense on the part of the nurse [were] essential, for without the vision of the importance of their task they could not long endure the endless stair-climbing, the weight of the bag, and the pulls on their emotions."[33]

Although the nursing service retained its original aim to care for the sick and to solve related social and economic problems, growth required the creation of an institutional structure that could ensure consistency in all programs throughout the city. Inevitably, the morning family meetings at the Henry Street Settlement were replaced by an expensive and complicated system of communication and administration: board of directors, committee on nursing, supervisors, rules, "Bulletin of Instruction," and extensive "carefully kept" records.[34]

Wald's decision to incorporate her organization required many of these bureaucratic changes. Although Jacob Schiff saw no need for incorporation, Wald thought that, as a legal entity, the settlement was in a better position to amass a large sustaining fund. Corporatizing and institutionalizing the work might change the character of the organization, but the ever-present anxiety associated with an intermittent and unpredictable income, and the risks for Wald of passing funds through her personal account, overshadowed these concerns. The newly appointed board immediately created the Visiting Nurse Permanent Fund, which with assistance from Schiff's investment firm held assets valued at $502,000 and produced an annual income of $25,000 in 1916.[35]

The nursing service grew through the 1920s, with eighteen nursing centers throughout New York City staffed by 164 field nurses and twenty-eight supervisors. By 1926, the service cared for over 49,000 patients and made nearly 347,000 home visits. Prenatal and preschool clinics, mothers' clubs, and well-baby consultations provided 10,567 hours of clinic care for 18,330 patients. Almost 50 percent of the nurses' visits were in connection with the maternity program, double the proportion of such cases in the previous decade. Half of the patients were under five years of age, over 70 percent were women. Even with declines in immigration in the 1920s, one-quarter of the settlement's work focused on "Americanizing our foreign-born population" (Italians, 28%; Russians, 18%; Irish, 14%; and Austrians, 13%) through health teaching.[36]

Henry Street nurses began caring for the black community in 1906, when Wald was visited by a "race woman" who challenged her to do something for the black community. Wald accepted the challenge, hired the woman (Elizabeth Tyler), and found the funds to set up a branch of the settlement in a small storefront in the San Juan Hill (now Columbia Hill) section of the city. Wald had no problem finding "admirably trained and efficient" black nurses, and the positive community response to the settlement's social, health, and educational programs was immediate. By 1925, 15 percent of the settlement's patients were black. With the death rate among blacks twice that of the rest of the city's population, the black community received "special attention" as the black nursing staff increased from one supervisor and four nurses to two supervisors and eighteen nurses in the 1920s.[37]

Wald Retires

In 1933, after forty years of service, Lillian Wald retired as head of the Henry Street Settlement. She ended her career as it began, confronted by economic depression. But each year her staff of 265 now climbed 24,750 sets of stairs, drove 140,000 miles, and took 120,000 subway rides to make 550,000 home visits to 100,000 patients. The nurses cared for one-fourth of New York's cases of pneumonia, one-third of all maternity cases, and one-fifth of reportable diseases. The nurses were to the hospital system what the extension course was to the university. The plea "Send a nurse!" still came with assumptions linking nursing, social welfare, and the public, but the response was much harder to finance now than when Wald moved to the lower East Side in 1893.

At the time of her retirement, Wald and the board wanted to preserve the original structure of the Henry Street Settlement with its interplay between social services and nursing care and to retain a single head for all programs.[38] In the end, however, the united programs of the settlement simply could not be sustained in the absence of Wald's leadership. The original vision of an "organic structure" uniting social services and nursing ended. Lillian Wald died in 1940 at her home in Westport, Connecticut, after a long illness. In 1944, the social and nursing activities of the settlement were separated. The nursing program moved to its headquarters farther uptown, becoming the present-day Visiting Nurse Service of New York, while the Henry Street Settlement remained focused on the neighborhoods of the lower East Side.[39]

But, as Wald concluded in 1934, "No organization proceeds far with a fixed idea." She simply hoped that "the day may soon dawn when we Americans can enjoy a measure of life and health that is consistent with our extraordinary resources and the intelligence of our people. The pioneers have begun the work; it is far from finished. New fields, new enterprises, are visible. The times call for the high spirit of the courageous pioneers among physicians, scientists, and nurses."[40]

Home Nursing Care—Yesterday, Today, and Tomorrow

$\mathcal{P}$hotographs help convey the meaning of home care for providers, patients, and family members. Historic photographs of home care are located primarily in the collections of urban visiting nurse associations. Most of these images were used in annual reports to dramatize the importance of the nurses' work, with the hope of increasing financial donations from the VNA's supporters. It is important to remember that each image was chosen to convey a specific message.

The Nurse's Work

Most early visiting nurses made between eight and a dozen home visits each day. A typical day began at 8:00 a.m. with visits to patients who were considered seriously ill and concluded eight or ten hours later with a return visit to the same patients. Nurses devoted the rest of the day to the care of mothers and their newborn babies, patients with tuberculosis, injuries, abscesses, and fevers of various origins, and a few with chronic conditions. The nurse generally stayed in each home for half an hour to two hours, depending on the severity of the patient's problem.

Nurse caring for a patient's infected foot, circa 1912–14. *Visiting Nurse Association of Chicago, Chicago Historical Society.*

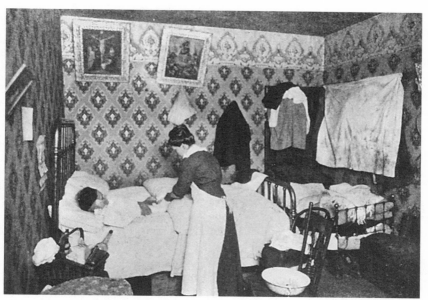

Nurses often cared for very sick patients at home. The original caption described this as a "case of peritonitis and festered wound which has never healed. Family lives in one room"; 1911. *Visiting Nurse Society of Philadelphia, Center for the Study of the History of Nursing, School of Nursing, University of Pennsylvania.*

The visiting nurse's mission often included guiding immigrants in the "American way." The original caption for this photograph described the nurse as "instructing Italian mothers as to the care of their babies"; 1912. *Visiting Nurse Society of Philadelphia, Center for the Study of the History of Nursing, School of Nursing, University of Pennsylvania.*

Getting There

Early visiting nurses usually got to their patients' homes on foot, often walking ten miles and climbing endless numbers of tenement stairs daily. Reaching their patients was rarely a simple undertaking. Over time, walking was supplemented by bicycles, carriages, trolleys, and eventually automobiles.

"Over the rooftops" between patients in the tenements in New York's lower East Side, circa 1910. *Photograph by Jessie Tarbox Beals, Henry Street Settlement, Visiting Nurse Service of New York, Center for the Study of the History of Nursing, School of Nursing, University of Pennsylvania.*

Patients

Photographs were used to illustrate the depth of poverty encountered by these nurses. Typically patients' homes were portrayed as filthy, small, and crowded. A picture of a family would frequently be accompanied by a caption describing how father, mother, and seven children lived in two rooms that they shared with four boarders. Patients were characterized as sad, needy, alone, and old as well as sick, and they were often characterized as immigrants or "colored." Visiting nurses were depicted as welcome, expected, and frequent visitors.

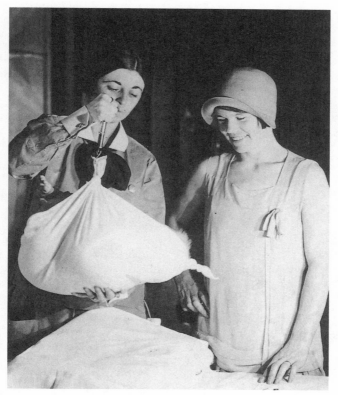

Infant wiggles while being weighed by visiting nurse, 1920s. *Visiting Nurse Service of New York, Center for the Study of the History of Nursing, School of Nursing, University of Pennsylvania.*

A chronically ill patient expectantly awaits "her nurse's" visit—no matter the weather; 1951. *MCP Hahnemann University, Archives and Special Collections.*

Still Welcome Visitors

Today's "crisis in caring" is characterized by new, frightening diseases and the burdens associated with chronic illness, rising health care costs, a confusing health care system, and, for many families, unmet health care needs. Contemporary photographs remind us that nurses remain a practical solution to some of our most urgent health care problems.

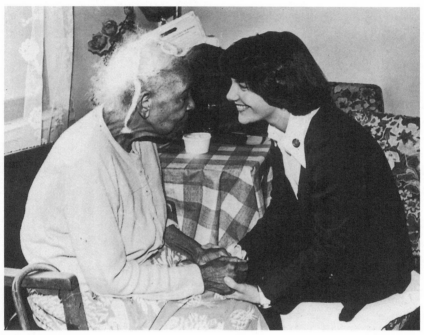

Successful home care has always required a combination of caring, trust, and competence; 1970s. *Boston Visiting Nurse Association, Boston VNA/IDNA/CHA Collection, Department of Special Collections, Boston University.*

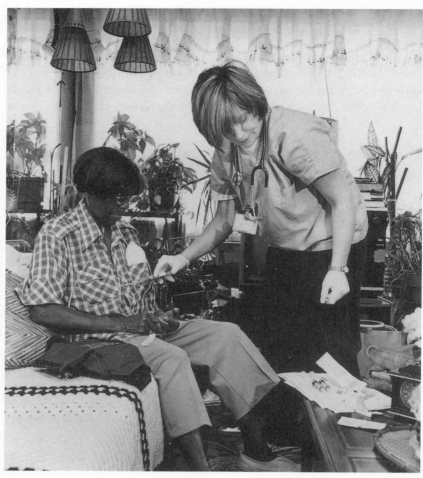

Since the turn of the century, visiting nurses have frequently introduced advancements in technology into their patients' homes; 2000. *Visiting Nurse Association of Greater Philadelphia, Center for the Study of the History of Nursing, School of Nursing, University of Pennsylvania.*

Part III

Management and Money

CHAPTER 6

The Business of Private Nursing

*B*y the beginning of the twentieth century, more hospitals were being built but most sick people, rich or poor, were sick at home. With work in hospitals rarely available, nurses were primarily employed privately, in patients' homes. These nurses included trained nurses—expert replacements for family caregivers—and "self-styled" nurses, who lacked formal training. There were fewer than five hundred trained nurses in the United States in 1890; inevitably, the purchase of their services was limited to the more affluent. Furthermore, most families and physicians remained unconvinced that the sick at home required a trained nurse. As pupil nurses, untrained nurses, domestic servants, and "unclassified" help were pressed into service, trained nurses increasingly found themselves in competition for employment.[1]

"Grit, Gumption and Grace"

Private nursing embodied all the complexities of care at the turn of the century. Nursing, as one nurse described in 1915, was "science and its instrumental art of application . . . practice and production."[2] Although the young nurse was said to leave the hospital loaded with etiquette and theory and "stupid with rules," this was not the sort of knowledge that would make her successful in patients' homes. To be practicable, hospital methods had to be modified and adjusted to home conditions. The value of the nurse was not in idealizing, theorizing, or using technical skills but in acting as "a kind of

125

go-between—a balance wheel, so to speak, in a family disorganized by sickness." In the words of a nurse from Norfolk, Virginia:

> We find, too, that to dive into the mysteries of human nature, a study for which we find ample opportunity, becomes almost imperative. As we enter a home, no matter whether it happens to be a mansion or a hut, it is, as a rule, to our advantage to observe the conditions, the circumstances, the atmosphere, the attitude of the people of that home. It is just as important to ascertain the patient's state of mind and temperament as it is to minister to the needs and comforts of the body. The patient's family occasionally proves to be a most trying factor to deal with, and even the servants, if there are any, come in for consideration. Besides the attending physician must be given evidence that we are worthy of his confidence. In coming into the midst of a household we ought to use much tact and discretion, so as to avoid giving the impression that we are a foreign element, a stranger with strange ways. Rather, we should [try] . . . unobtrusively to bring the home conditions back as far as possible under the circumstances to a normal state, though this is sometimes a difficult task to accomplish under adverse conditions.[3]

Tact, sympathy, courtesy, a sense of humor, the ingenuity to make do, superhuman self-control, and a spirit of helping out made a good home nurse. In her article "One Hundred Don'ts for Nurses," Jeannette Forrest outlined the frequently repeated list that included squeaky shoes, rustling uniforms, gossip, and talking too much. Number 100 on the list was an admonishment to the reader not to be discouraged if the list reminded her of mistakes— just "try again."[4]

Nurses employed in private homes were all too familiar with the saying "what can't be cured must be endured." What had to be endured included everything from the idiosyncrasies of patients, family members, friends, and servants to exhaustion, unlimited hours of "duty," isolation from the outside world, a nomadic life style, absence of home life, vagaries of taste, and "class problems"—living like the other half lives. Families ran the gamut from pleasant and capable of providing for the nurse's and patient's every need (no matter their resources) to "odoriferous," "ungrateful," and densely "ignorant." For the nurse in private practice, each new case simply presented a problem to be solved.[5] Nurses also had to cope with a variety of home conditions, including low beds, contrived refrigerators, and vermin—to name but a few. But the ultimate challenge was how to respond politely to the query, "Do you think we pay you to sleep?" As one southern nurse summed it up: the requisite qualities of a good private nurse were "grit, gumption and grace."[6]

Private Nursing and Typhoid Fever

Private nursing of the typhoid patient provides a classic case study of the work and business of nursing at the turn of the century. Given the menace presented by typhoid, it was little wonder that nurse-expert Janet Geister claimed that "typhoid fever nursing . . . was perhaps the most regular event" in the life of a private-duty nurse.[7] Typhoid was so life threatening and complicated that most patients could not get along without care. As one physician observed, "there [is] no condition in medical practice, infectious or not, in which the probability of employment of a trained nurse is so great . . . Families simply [cannot] manage such demanding care on their own."[8]

Typhoid was also a disease for which the outcome was said to depend more on good nursing than on medical care. Even the great turn-of-the-century physician William Osler acknowledged that typhoid fever was not a disease treated mainly with drugs. Rather, careful nursing, a proper diet, and hydrotherapy were essential to recovery. In the treatment of typhoid, he declared, an intelligent nurse should be in charge. Given the relentless nature of typhoid, its complications, and its potential for sudden death, care of the patient covered the whole gamut of nursing knowledge. Yet, as one nurse noted, "in no other disease may the patient be reduced so low with so much chance for recovery."[9]

Intelligent nursing of the typhoid patient was also critical because of the need to prevent spread of the disease to the nurse ("morbus Medicorum"), family, and community. As a classic food- and water-borne disease, typhoid was transmitted via flies, fingers, food, feces, and fomites (shared inanimate objects such as contaminated sheets). Every case implied a previous case. It often fell to the nurse to locate the source of infection. As early as 1890, articles in the journal *Trained Nurse* warned nurses of a moral obligation to protect the community through necessary prophylactic measures. Although patients were virtually quarantined by physical weakness, their disease could and did move on.[10]

If a nurse agreed to take a typhoid case, she was expected to report to duty at once, taking sufficient clothing, medical supplies, appliances, and reference books for an indefinite period (Figure 12). With hurried preparations

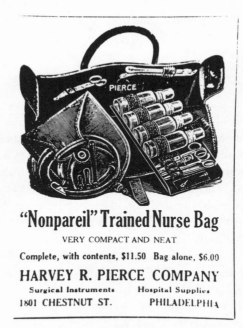

"Nonpareil" Trained Nurse Bag

VERY COMPACT AND NEAT

Complete, with contents, $11.50   Bag alone, $6.00

HARVEY R. PIERCE COMPANY

Surgical Instruments   Hospital Supplies

1801 CHESTNUT ST.   PHILADELPHIA

FIGURE 12. Advertisement for a Nurse's Bag from the *Directory of Trained Nurses* for Philadelphia, New York, and Brooklyn, 1895. *Private collection of Joan Lynaugh.*

and verification of instructions from her registry, the nurse could expect many weeks of providing critical care in the home of strangers. The reality of the private practice of nursing at the turn of the century was well expressed by one nurse who, confronted by terrible weather and delayed trains, descended alone into the rain, mud, and darkness of a strange town to find no one to meet her. She reminded herself that nothing but duty would cause her to expose herself to such a situation. As she would later write, who but a nurse knew such feelings of depression—homesickness mingled with forebodings of what awaited? One nurse, having survived the ordeals of a typhoid epidemic in a small Pennsylvania town, suggested that to her class motto, "Deus et officium" (God and duty), should be added the phrase "Qui patitur vincit" (He conquers who endures).[11]

Not uncommonly, nurses found living conditions considerably less than desirable. Often unable to leave the typhoid patient's bedside, many nurses interrupted their "lonely watch" only to sleep for brief intervals. One nurse described the dreary hours of the night, cheered only by the musical snores of family members. A creaky cane rocker without footstool or pillow was her only creature comfort. "Between frequent orders, a feverish patient, snoring mother, and poor facilities for ventilation the hours passed by bringing the welcome day light." Rarely did these nurses have the luxury of a decent night's sleep, since rest came only when patients were stable or family members could provide relief. "Fair" beds and dirty quilts meant that nurses usually slept in their clothes. One nurse described brief moments of sheer bliss—lying down after a bath, undressed, on a sheet on the hard wood floor in order to relax for a few minutes each day. Admitting that she could never have supposed this would give such a sensation of luxurious ease and great refreshment, she concluded that comfort was without doubt a relative quality.[12]

Adequate bathing proved as difficult as sleeping. Often the only available facility was the kitchen faucet and a most unsatisfactory "family towel." One nurse "enjoyed" a bath in a tumbler of water, which she considered an improvement over no bath at all. In her diary she wrote, "How fortunate I came here perfectly rested! How long can I stand nursing like this? How long will my love for the work or the money earned seem compensation enough for all these inconveniences?"[13]

Beyond the numerous inconveniences, the complexity of caring for such desperately ill and highly contagious patients called on the nurse's every ability. An intelligent understanding of the nature and progress of this "great" fever was essential. Given the "facts" about typhoid at the turn of the century, care centered around the creation of "perfect rest," fever management, proper nutrition, prevention of complications associated with prolonged bed rest (bedsores, wasting, and pneumonia), administration of medications and treatments ordered by the physician, and management of emergencies associated with hemorrhage and intestinal perforation, the most common causes of death.[14]

The onset of typhoid fever was insidious, with general malaise followed by nausea, vomiting, and diarrhea. At this early stage of illness, families who could afford it commonly sought the services of a private-duty nurse. During the first week, the patient's fever steadily increased to dangerously high levels, typically reaching 103 to 105 °F, often accompanied by severe headaches and delirium. Constipation or diarrhea also often began during the first week, along with cough and bronchitic symptoms. Tongue, gums, and lips could be coated with sores, and some patients experienced abdominal symptoms of tenderness and slight distension. The appearance of characteristic rose-colored spots confirmed the diagnosis.

The discomfort caused by high fever, rapid pulse, muscle weakness, diarrhea, tympanites (abdominal distension caused by gas), and abdominal tenderness became exaggerated during the second week along with worsening mental status. In more severe cases, death occurred at this stage, often associated with pronounced "nervous symptoms," hemorrhage, or intestinal perforation. In the persistent typhoid state of the third week, the pulse was high, temperature showed some morning remission, and fever began to decrease. The patient was very weak, and pulmonary and cardiac complications were common. Delirium was often pronounced and the danger of intestinal hemorrhage and perforation remained high. Convulsions sometimes occurred during the fourth week, but a continued decline in temperature and lessening of diarrhea were also common; appetite usually began to return. Weight loss, anemia, and fatigue were significant during this stage, and in severe cases the patient became very weak. Typically, the patient's pulse was rapid and feeble, tongue dry, and abdomen distended; a profound stupor with "low muttering" delirium was also common. Death at this stage resulted from circulatory failure and complications. For patient, family, and caregiver, typhoid was a wretched disease.[15]

With this or some similar trajectory of illness, private nursing emphasized, among other things, bed rest and skin care. Activities were structured to conserve the patient's energy and required careful planning. A daily bath was the rule, but attention to a patient's skin after each episode of diarrhea was also essential, and these episodes could occur as often as ten times a day. The bath was usually completed by rubbing the patient from head to foot with equal parts of alcohol and water. When available, clean bed linens were provided. Of special concern was the prevention of bedsores through cleanliness and frequent changes in the patient's position, all performed without fatiguing the patient. Mouthwashes and swabbing of the teeth with a mild antiseptic solution or a mixture of water and borax, lemon juice, glycerine, or claret were used to treat the effects of typhoid on the tongue and mouth. Nourishment most often came in the form of milk. The patient's tolerance and preference dictated whether it was simply boiled, diluted, or given in "modified" forms such as peptonized, fermented, or malted milk, buttermilk, or whey. The nurse devoted considerable energy to finding a diet suitable for the patient. Milk was flavored to disguise its taste, and oyster or beef broth, junket, raw eggs, gruels, and jellies were gradually added to the daily diet. Small frequent feedings were the usual approach, with some physicians insisting that the patient be awakened at intervals throughout the night for nourishment.[16]

Hydrotherapy, "the use of water, inside and out," was supposed to subdue the fever and sustain the patient. Commonly, when the patient's temperature reached 102.5 °F, hydrotherapy treatments were done every third hour until the fever abated. Hospitalized patients could be treated with "plunge" tub baths, but the four people needed to lift the patient were rarely available in the home. In private practice, sponging and the use of a wet pack were the usual methods for reducing fever. One nurse wrote in the *Trained Nurse* that nurses should know why they sponged and should endeavor to become an artist in sponging. Lowering the temperature was of secondary importance to the effect on the nervous system—quieting delirium, assuaging restlessness, lessening insomnia, toning the nervous system, steadying the pulse, and producing tranquility. Because sponging the patient was frequent and necessary, the nurse had to ensure that the patient's response was pleasure not dread. Sponging required long, firm, straight, downward strokes, paying particular attention to the large blood vessels and to the spine—touch should be gentle, firm, and soothing.[17]

Tympanites was treated with both heat and cold, including "turpentine stupes" (cloths soaked in hot turpentine), poultices, an ice-coil, and rectal tubes. Each treatment required special preparation to protect the patient's skin. The use of turpentine, for example, was preceded by covering the abdomen with a light coating of petroleum jelly so the heat could be tolerated and blistering avoided. Enemas were also used to reduce tympanites, as was gentle massage twice a day.

Hemorrhage and perforation, the "accidents" of the febrile stage, called for great skill and judgment. Hemorrhage could be "concealed or open," which meant that death could occur before any blood appeared in the stool. Therefore, between the end of the second and beginning of the fourth week, the nurse was gravely concerned when her patient appeared pale or mentally excited, experienced a sensation of sinking, and had a rapid, weak pulse. The standard response was to initiate absolute bed rest—keeping the patient "rigidly quiet." Food was restricted, ice applied to the abdomen, and the foot of the bed elevated. Intestinal perforation was most often marked by sudden, acute abdominal pain of increasing severity; collapse; a rapid and weak pulse; cold, pale, and moist skin; marked rigidity and tenderness; and

abdominal distension. Hiccoughing was a common and disturbing response. Because perforation was often fatal without surgery, early diagnosis
and summoning of the physician were the nurse's most immediate tasks.[18]

Although Osler had suggested that typhoid fever was not a disease to be
treated mainly with drugs, contemporary practitioners' uses of such treatment were "numerous and popular." No single drug gained prominence
during this period; each was thought to perhaps lessen complications or
slightly shorten the duration of fever. Whatever their effectiveness, they remained part of the care and were a time-consuming activity for both patient
and nurse. Not surprisingly, changes in doctors often resulted in changes in
medications and treatments.[19]

Finally, the nurse had to face the all-consuming matter of preventing the
spread of typhoid to others. The outcome of prophylaxis rested fundamen-
tally with the nurse's management of the patient and his or her "dejecta."
Handling any article that had come in contact with the patient, whether
clothing, linens, dishes, eating utensils, bathwater, or bedpans, required
great care. The danger of spreading typhoid called for the nurse to be "overly
careful and morbidly conscientious." Although the rules for disinfection
were generally practical (one nurse reminded her colleagues, "Germs are not
athletic animals able to leap to heights . . . they are subject to the laws of
gravity") and the dreaded outcome of failure to comply widely understood,
the task was nevertheless formidable. *Salmonella typhi* was extremely hardy.
Feces, urine, vomitus, and bathwater were treated with chlorinated lime or
formalin and allowed to stand for one to two hours. Bedpans and urinals
were boiled for thirty minutes; linens and bathwater were treated with formalin and allowed to stand for ten hours. Patients and the public were protected from flies and mosquitoes by mosquito netting attached to the corners of the patient's bed, with an opening on each side that allowed the nurse
to give care. Fundamental to all these procedures was the nurse's rigorous
attention to disinfection of her hands and rubber gloves, if they were used.

These efforts at avoiding the spread of typhoid were often being undertaken in households where the family had absolutely no understanding of
the nature of the infection beyond a passing knowledge that typhoid was
"catching." On occasion families registered complaints, with accusations
that all these efforts were no more than "high-toned foolery." One nurse,
confronted by what she described as an ignorant family's resistance, wrote
of conjuring up "memories of Clara Weeks, or was it Isabel Hampton, or was
it the stereotyped answer we knew would be on examination papers?" in her

search for a practical method appropriate to a farmhouse void of all conveniences—but located next to the community's water reservoir.[20]

The Business of Private Nursing

According to Lavinia Dock, the *real* challenge in private nursing was for patient and nurse to find each other. Success required maneuvering among various "local" arrangements but often depended on chance. Some hospitals maintained directories of registered graduates. Nurses could also receive cases directly from physicians, hoping eventually to build up a reputation as well as a dependable practice. Newcomers and nurses unattached to hospitals with registries resorted to listing their names in city directories. Here, they were often indistinguishable from "nurses" of all descriptions.[21] In most cities, local directories were maintained by drug stores, commercial registries, or occasionally medical societies. Physicians found recommending or procuring nurses for their patients a time-consuming responsibility. Before the days of telephones or taxicabs, obtaining a nurse "at the moment of sickness or accident" was a difficult undertaking requiring a messenger, family member, or the physician to travel miles. Even then, the search or subsequent arrangements might prove unsatisfactory. Families found the process equally "clumsy and unwieldy." Lavinia Dock described the process of finding a nurse as not a "systematic, expeditious and sure service" but often a "vast annoyance," with nurses waiting at home for work while doctors and family spent hours in the search for a nurse. These aggravations notwithstanding, Janet Geister, the champion of private-duty nursing, claimed in 1910 that it was a field giving "the majority of its members genuine satisfactions, a reasonable money return, and a steady job."[22]

A circular sent to physicians in Washington, D.C., in 1882 outlined the problem and the standard solution:

> Even if a list of nurses be kept it is impossible to know who is engaged and who is disengaged, and hence, not infrequently, much valuable time is lost in anxious and fruitless search over all parts of the city before one can be obtained. At the same moment many excellent nurses are idle, who, if the fact were known, would be instantly employed. What is needed is some way to bring the two classes together—a Directory which will keep a list of all the competent nurses, both male and female, who wish to register, with an accurate record of their addresses, qualifications, charges, engagements, etc., so that physicians and others may be able to procure desirable nurses for all classes of patients, without delay.[23]

A central directory was a fairly obvious and logical solution to the individualistic, inefficient, and chaotic methods of distribution prevailing in most cities. Success, however, required that doctors, nurses, and patients agree to the "hearty support" of a single directory. As early as the 1880s, contention about directories arose over issues such as control, eligibility for registration, distribution of work, rates, and "profits." The ensuing competition among hospitals, doctors, nurses, and commercial registries to create "a great dependable utility established and conducted in a pure spirit of altruism" produced predictable charges of commercialism, unionism, extortion, favoritism, price fixing, quacks, blacklists, and failure to protect the public.[24]

Nursing journals of the period recorded these debates and rivalries, as did the popular press on occasion. In 1895, Lavinia Dock introduced the idea of a central directory owned and controlled by nurses. She described it as an ideal system of distribution, offering the public "a high state of perfection" and nurses "due attention." Dock's call, and the calls of others, to "unite" to "own themselves," to become "a highly organized branch of service, governed by its own codes, pruned of unworthy members by its votes, and managed as to its business affairs by its own representatives," went unheeded. Some "observers" saw exclusive central registries as a "curious egotism," and attempts by advocates of central registries failed to produce the organizational infrastructure required for union. The debate over a central directory continued for twenty-five years, even as communities developed a vast array of locally acceptable methods for distributing private nurses.[25]

In the late 1920s, only 24 percent of nurse registries were centralized directories. Most (63%) were controlled by hospitals or training school alumnae associations, and the rest were operated commercially by firms or individuals. Registries were unevenly distributed geographically, with most located in the North Atlantic and North Central cities with populations greater than twenty-five thousand. In the South, West, and rural parts of the country, families cared for the sick at home as best they could. In the meantime, nurses turned increasingly to registries for work, intensifying the uneven distributions of nurses in private practice. Of the three hundred thousand nurses in 1920, about half were trained and registered. With an estimated 120,000 nurses working in private duty, the ratio was two nurses (one trained and one untrained) to every 294 people. The number of nurses exceeded the available work; the field of private nursing was clearly overcrowded. While families and physicians were baffled by the abundance of and options for nursing care, nurses were bewildered by longer waits for cases and sharp fluctuations in the availability of work.[26]

Even as the process for "putting the right nurse in the right place" became more competitive, fragmented, and irrational, the business of nursing nevertheless persevered.[27] The registry, in its many forms and irrespective of ownership, remained fundamental to the placement of nurses among purchasers of care for the sick at home. For families, the task of finding a nurse most commonly began with a request to the physician for guidance or procurement. If left to their own resources, families accustomed to finding household help would begin with the city directory or classified advertisements in newspapers, where they could find several indistinguishable organizations listed as Nurses' Directories.[28]

Many city directories listed the names of individual nurses. Some of these listings might simply include only a name and address, whereas others indicated the nurse's training and title (e.g., nurse, trained, graduate, registered, or midwife). If not evident, male gender was specified. The Philadelphia directory listed name, address, and neighborhood only, under the heading "Nurses" in its business section. With such limited information, it is hard to imagine how prospective customers selected from a list of two hundred indistinguishable nurses. Obtaining a nurse from the Philadelphia College of Physicians must have seemed a much simpler and more prudent strategy. The college listed its registered nurses and phone numbers in an advertisement in the shopping guide section of the Philadelphia city directory. It promised a skilled nurse for all classes of patients at the shortest notice. It is little wonder that the College of Physicians directory was both successful and profitable.[29]

In southern city directories, while training might not be obvious, race was always apparent. In Charleston, South Carolina, for example, *c* identified the nurse as black (colored). Use of *Miss* or *Mrs.* identified nurses as white, the lack of an honorific indicating the nurse was black. In the North, city directories did not make these distinctions, but racism took other forms. Claiming they had no calls for "colored nurses," northern registries "politely" declined requests from black nurses to be listed. In her autobiography, Jane Edna Hunter described the deep prejudice she encountered following a move from Charleston to Cleveland, Ohio. In Cleveland, the "rebuffs" were much more severe than those experienced in the South. One northern physician even told her to go back to the South—"White doctors do not employ nigger nurses" in Cleveland. Hunter placed advertisements and visited physicians' offices in a desperate search for work, but she found even "un-prejudiced" physicians to be unresponsive. In despair, she resorted to cleaning and laundry jobs, until she had the "good luck to wander into the office of Dr. L. E. Sieglestein" and finally obtain employment as a nurse.[30]

The Impact of the Telephone

By the first decade of the twentieth century, city directories, which at first listed nurses by title and address, also included telephone numbers. Registries were among the earliest businesses to acquire phone services, and most also purchased long-distance connections. Telephones instantly transformed the ability of families and physicians to find a nurse. Families could usually avail themselves of a telephone at the nearest drugstore. Nurses with telephone access had a great competitive advantage. In some communities, private nurses subscribed to more than one phone service so they could be reached by families within the city and in smaller towns outside the city. As the telephone absorbed a considerable portion of registries' expenses, nurses without phones were frequently charged a fee if messenger service was required.[31]

The advantages of being available by phone were readily apparent. The records of a Philadelphia alumnae association registry show that in 1904 most calls for nurses went to those with telephones or those living very short distances from the registry office. The cost of telephone services was exorbitant, so nurses in boarding houses had the advantage of sharing the expense of a communal phone. In the South, trained white nurses were more likely than trained black nurses to enjoy the advantages of being available by phone. By 1922, approximately 40 percent of black nurses and 67 percent of white nurses listed in Charleston's city directory were reachable by phone.[32]

The telephone has been described as a technology with "effects in diametrically opposite directions." It saved physicians from making house calls, while at the same time allowing families to summon the doctor at will.[33] The steady increase in residential telephone subscribers also made it easier to find a nurse, obtain medical help in emergencies, and seek routine caregiving advice. As families' isolation was reduced, along with the sense of imminent peril, a private nurse seemed less essential than in the past. Contact from friends and neighbors now came in the form of calls offering words of encouragement, rather than physical presence and help with caregiving. Finally, while registries and nurses could more easily stay in touch, easy access also produced conflict. Nurses could call the registry on any pretext, from reporting a half-hour outing to announcing plans to attend church. Concerned that these frequent calls tied up the phone lines, registrars saw calls to ask "Where am I on the list?" as particularly "thoughtless." One desperate registrar even described the need for "a telephone operator just to receive the calls from nurses alone." Likewise, nurses reported a number of "telephone incidents" in which a registrar's disagreeable actions, manner, or attempted discipline caused offense and hurt feelings.[34]

Those in search of a nurse in Philadelphia could also refer to publications such as *Cornell and Shober's Directory of Trained Nurses of Greater Philadelphia for 1900,* or Miss Longeway's *Directory for Nurses.* They might even check with the local drug store for lists of "off duty" nurses or consult classified advertisements placed by nurses in local papers. Whatever the strategy, families disorganized by sickness had to select a nurse from among strangers and allow "a foreign element" into their household.[35]

The Registry: Bringing Nurses and People Together

Commonly, registries served as the intermediary among physicians, nurses, and the public. As a speaker reminded her audience at a 1909 nursing convention, registries succeeded because, like any good business, they could supply "the demand for a certain line of goods quickly, in greater numbers, in greater variety, and cheaper than the individual can procure the same goods for himself."[36] *Quickly* meant that the nurse was prepared to respond to calls within thirty minutes. Greater numbers of nurses were needed by the registries, as they increasingly allowed nurses to "register against certain classes of work," such as (to use then current terminology) insane, colored, hospitals, tuberculosis, men in bachelor apartments, contagion, out of town, and cases after 9 p.m., or to "specialize," for example in obstetrics. Nurses often registered with several directories and then resigned from them once their reputation was established in their registry of choice. Such withdrawals of apparent loyalty and moral and financial support were seen as a "great mistake" and detrimental to a registry's prestige and ability to succeed. At the turn of the century, economic viability required that registries maintain a list of at least one hundred nurses, but until the late 1920s, even larger directories had difficulty filling requests for nurses.[37]

All registries required payment from nurses wishing to be listed, $2 to $3 per year before 1900 and $10 to $12 by the 1920s. Some "commercial" registries charged nurses a percentage of their earnings, usually 10 percent, as well as charging families for the use of the registry. In 1882, for example, the Washington, D.C., directory for nurses charged those interested in obtaining the addresses of "disengaged" nurses $1 between 7 a.m. and 10 p.m., $2 if this information was needed after 10 p.m., and an additional $1 for "finding and sending" a nurse.[38] In the 1880s, graduate nurses received $15 to $18 per week; by the late 1890s, the rates had increased to $20 to $25; and by the late 1920s private nurses were charging $40 to $50 a week. Higher rates were charged for so-called mental, contagious, and alcoholic cases. Although rates varied regionally, the cost of private-duty nurses meant that only under the direst of circumstances were they found in more humble households.[39]

Matching nurse to patient and doctor was the complicated responsibility of the registrar. While most registries were open from 8 a.m. to 8 p.m., registrars were expected to remain available twenty-four hours a day, seven days a week. The selection of a registrar was of paramount importance, for this was the person with whom physicians and families interacted. Satisfactory service was largely dependent upon the registrar's interpersonal skills. Like the nurses, she was expected to bring almost superhuman talents to the job. "She should be a woman of education and business ability, possessed of a sympathetic appreciation of the aims and ideals animating the medical and nursing professions, and should be versed in professional ethics, for without the latter qualifications she will be unable to recognize and to handle with tact, keen judgement and dispatch the many ethical situations that will require this."[40]

The success of the registrar and registry required keeping track of hundreds of details, organized in some quickly accessible system. One specimen of nursing ingenuity employed a 12.5-by-16 inch "board" to organize a list of 170 nurses (Figure 13). The board contained forty hooks arranged in rows of ten, numbered consecutively. Each nurse had a small card with name, address, and "telephone call," which was placed at the bottom of the list when the nurse became available for work. In the upper left corner were listed, in abbreviated form, any cases the nurse was unwilling to accept, such as "No Con., Tbc, Ob, Out of City, male, 9 p.m." Cards also indicated specialized interests such as massage or obstetrics. Varying colors on the cards indicated where the nurse trained; green indicated "foreign nurses" from out-of-town schools. The cards of nurses out on cases were arranged alphabetically in metal slots on the back of the board. Nurses available for work were expected

FIGURE 13. Registry Board Used by the Graduate Association Nurses'
Registry in Minneapolis, 1910. *American Journal of Nursing.*

to be reachable and to keep the registry informed of their whereabouts. A tag marked with their location was then placed over their card. A larger tag was used when nurses made an engagement for a definite date. A tag was also used for nurses returning from "cases of contagion," to allow physicians to decide the advisability of placing such a nurse on special cases. A ledger book was used to record doctors' orders and a journal to record details of the nurses' cases (date and time, in and out; doctor; patient, address, disease, and remarks).

When a request was made for a nurse, the registrar inquired about the nature and location of the case and any preferences for a certain nurse or school. If no preferences were expressed, the registrar called the nurse uppermost on the board. If the caller requested a specific nurse, and if she was available, that nurse would be provided no matter what her position in the queue. If the caller requested a graduate from a specific school, the registrar, using her color-coded card system, would read the names of nurses from the designated hospital. The same procedure would be followed for specific problems, such as contagion, with the registrar proceeding down the list until a nurse matching the request could be found. A nurse's refusal of any call, unless it was a class of work that she had registered as "not taking," placed her "at the foot of the list."[41]

How Did She Work Out?

When a family had finally found a nurse, family and nurse were left to manage at home alone except for the physician's periodic visits. There were no methods of oversight or supervision for nurses, much less physicians, employed in patients' homes. Nor were there any regulations to protect patients and families. Anyone could open a registry and employ any sort of nurse. Nurses who, in the common parlance of the day, had "failed to make good," and even those with histories of poor work or conduct, drug addiction, problem drinking, neglect of duty, or a history of making false reports, could eventually find a registry willing to add them to its list. Reports of stealing— jewelry, diamonds, money, and even husbands—occasionally appeared. Standards and quality control were totally dependent on the registrar's vigilance. Although registrars were empowered to remove nurses from the list for "sufficient cause," such extreme actions were rarely recorded.[42]

Registries sponsored by graduate nurse associations claimed a higher level of stewardship and oversight than the commercial registries. Screening of applicants' credentials included inspection of diplomas, letters of reference, an interview, and a photograph. Confidential performance evaluations by families and physicians, customarily obtained at the conclusion of each case, formed the basis for oversight. These evaluations asked for an appraisal of the nurse's conduct, efficiency, desirability, temperament, neatness, health, faults, mindfulness in following doctors' directions, tendencies to gossip, and agreeableness as a member of the household. Nurses thought their success was also dependent on their "ability to nurse," but this was not a usual category of inquiry.[43]

Indeed, wise nursing or "practical and theoretical perfection" was not articulated as the standard for employment or success. Writing to the editor of the *Trained Nurse* in 1894, the "proprietor" of the Alpha Nurses' Agency spelled out the other critical standard of the day: "to the best nurses from the physician's standpoint [goes] the most work." In other words, a physician's response when asked if he would employ the nurse again, and in what type of cases, determined a nurse's future employability. Nurses who worked regularly were those wise enough to develop good relationships with physicians.[44]

For private-duty nurses, one other complex question remained, "Who is in charge?" Was it the patient, nurse, or doctor? Signing his letter "The Grouch," one patient wrote to the *Trained Nurse* in 1915, wondering if the nurses he employed during a recent illness were engaged for his welfare or the doctor's. If he could tell his cook what he wanted, then he should be able to do the same with his nurse. He was perplexed to find that the nurse apparently felt no obligation to her patient other than to obey the doctor, even against the patient's wishes. The Grouch concluded, "Is it impossible for me to get a trained nurse to do *her* best for *me* or must I, if sick again, depend upon my cook and my parlor maid?" In answer, A.E.C. reminded the readers of the *Trained Nurse* that nurses were usually employed as the result of a doctor's suggestions and that, while "the nurse is undoubtedly employed by the patient; still she is *always* under the doctor's orders."[45]

Physicians certainly would have agreed. From their perspective, the nurse was the handmaiden of the physician, never an equal. Physicians were shocked by a nurse whose "faithfulness to the patient" caused her to "stand in" with and actually obey the patient rather than the physician. No matter where nurses publicly situated their loyalty, the reality of home care often meant that doctors' many long absences required nurses to become self-reliant and capable of making, in Janet Geister's words, "momentous decisions." Little value accrued in what Dock referred to as "obedience which is slavishness" or "subordination which is moral cowardice." A rather ambiguous autonomy, peculiar to circumstances of home-based care, dictated where the nurse placed her "duty and loyalty." Each situation was different. As Katherine DeWitt, author of a major book on private-duty nursing, explained, "It is like a game of chess—one can learn the rules of the game from another and can be taught how to make certain moves, or how to avoid certain complications that may arise, but after all, the player must manage the game in his own way and win and lose by his own ability. No two games are alike."[46]

"Fifty Years in Starch" and Fading Away

In 1938, Janet Geister wrote her classic article "Private Duty Nursing Then—and Now," beginning with the "good old days" when nurses enjoyed more "satisfaction, stability, and prestige." Then, private-duty nurses worked hard, slept little, and had no home life, but their dollars went farther and came in more steadily. Sometime between 1910 and 1915, everything changed. From Geister's perspective, what followed was a period "marked by revolutions so mighty that it will probably take twice that time for their significance to be understood." Overnight, science, telephones, good roads, motor cars, and new social attitudes changed modes of living and thinking. The individual practice of nurses and family doctors was at odds with the times. Advances in medical sciences radically changed the methods of treatment, as house calls and home care were replaced by hospital and office practice. In addition, patients' recovery time was dramatically reduced: the nurse who in the past packed enough belongings for a six-week stay now stayed two weeks at most. Patients sick at home rarely needed skilled nursing care, much less continuous care, and nurses found themselves waiting longer for cases as hospitals produced "a fresh deluge of graduates every year." The final blow was the appearance of visiting nurses in the homes of those in good circumstances, competing with the private-duty nurse "in her own field."[47]

Reaffirming Geister's appraisal, medical economist C. Rufus Rorem in 1933 described nursing's problem as an "economic paradox" of oversupply and underdemand. The records of most registries further illuminated this predicament. In Cleveland, for example, the number of nurses available through the Graduate Nurses' Registry increased from 250 to 2,000, while the Registry's ability to fill calls was dramatically transformed from 50 percent to essentially 100 percent between 1919 and 1926. Although nurses debated for two decades whether the field was becoming overcrowded, by 1928 the consensus was that there were too many nurses.[48]

The other significant alteration in private nursing was its movement from patients' homes to hospitals. By 1922, the United States had 4,978 hospitals.[49] Even in small towns, most medical care was centered in hospitals. With more of the middle and upper classes seeking hospital-based care, private-duty nurses predictably followed them into these institutions. For caregivers, the hospital was becoming an acceptable solution to the endless obligations of caring for the sick at home.

In Cleveland, for example, the first mention of a hospital call to the registry was in October 1909. As requests for private nurses began to come from hospitals, the registrar reported that few of the registry's members "care for institutional work." By 1910, 25 percent of calls were coming from hospitals for nurses to work as "specials" with individual patients. Some calls were also received for the less desirable job of hospital "general duty." By 1919, only 21 percent of cases were in patients' homes and the majority of private nurses worked in hospitals.

Within a decade, the preferences of private-duty nurses had shifted, and in 1920 concern was raised "in regard to private duty nurses' unwillingness to work outside of hospitals." This problem was so severe that in 1922 Cleveland's Registry Committee voted that nurses would not be allowed to "discriminate between hospital cases and home nursing." While the committee's chairwoman admitted it was "really unfortunate that we cannot have all our cases in hospitals," she was confident that the old question of "home duty" would not go away. "Our people still cling to their homes and still have a feeling that nurses are ministering angels." Nevertheless, home care was so unpopular with nurses that by 1926 only a small percentage of these calls could be filled.[50]

Sara Parsons, registrar of Boston's central directory (organized in 1912), reported similar transformations. By 1924, only 1 percent of calls were for home care, and of these about 40 percent "came from homes which it required grit" to enter. Parsons, who appreciated nurses' hesitancy about home care, reported being filled with awe and reverence when the nurses "tell me their struggles with bed bugs, and other disgusting problems." She claimed that not all these conditions were "confined to the homes of the poor, but exist sometimes in homes of feeble old people who cannot see and have not the strength to keep the house clean."[51]

Can You Send Me a Nurse?

By the mid-1920s, the work and business of nursing had entered a protracted transition from entrepreneurial home-based care to hospital-based private duty. Periods of oversupply and underdemand predictably meant irregular employment. The economic hardships created by the Great Depression concluded nursing's movement into the hospital. "Can you send me a nurse?" now had a very different meaning. By the late 1930s, most nurses worked in hospitals as private-duty rather than general-duty nurses. Only about 11 percent still worked in private homes. Local conditions shaped these practice patterns, with the demand for home care remaining higher in smaller towns. The nursing shortages during World War II and the early 1950s completed the transformation as hospitals increasingly hired nurses for general duty. Like most nurses, the 20 percent still working in private duty mostly worked in hospitals.[52]

Despite this recasting of nurses' place of work, an examination of city directories and want ads in local newspapers suggests that, however small, a market for home care still existed in the 1950s. As at the turn of the century, finding just the right nurse remained a challenge. Registries were now called "nursing directories." In Philadelphia during the 1950s, a few still had marginally informative names such as Stevens Nursing Directory, Visiting Nurse Society, Philadelphia Nursing Service for Christian Scientists, Ellen Phillips Nursing Directory, Nurses' Official Directory, Frances Directory, and the Nurses' League Corporation. Most promised to provide nurses day or night, male or female, and offered a range of credentials (graduate, registered, undergraduate, practical, or trained infant) and, presumably, prices. Families in need of a nurse could also examine several pages listing individual nurses for hire. These listings provided the nurse's name, address, and phone number. A small percentage also gave indications of preferences or specialty such as private-duty, prenatal, visiting, maternity, infants, aged, chronic, or convalescent.[53]

Newspapers also had advertisements for nurses. The *Philadelphia Inquirer* carried ads for both "help wanted/male and female" and "situations wanted/male and female." In addition to the information listed in city telephone directories, want ads included more personalized descriptions of the nurses such as kind, strong, no bad habits, reliable, steady, refined, resourceful, educated, not servant type, and experienced. Many mentioned a willingness to cook, clean, or sleep in. In these ads, those seeking help were most often elderly, semi-invalids, or new mothers and newborns. The nurses sought were either graduate or, more often, practical. Many ads listed either type as acceptable. Most interesting was the inclusion of information about race, by those seeking employment and those seeking assistance. Frequently included were descriptions such as colored, white, or light colored. Age preferences also appeared, suggesting a desire for mature, but not too old, workers. Salaries were described as reasonable or fair.

Despite the dominance of the hospital, even into the 1950s families employed nurses to provide home care. As always, what they sought was an idiosyncratic mix of services. Matching nurses in search of employment with families in need of their assistance remained a competitive, fragmented, and irrational system of distribution—but private home care survived.

CHAPTER 7

A Cautionary Tale

The Metropolitan Life Insurance Company's Home Care Experiment

*A*fter the turn of the century, the number of private-duty nurses and visiting nurses grew rapidly, the ranks of the latter somewhat irregularly as their organizations experienced variable financial support. Most visiting nurse associations (VNAs) were financed through donations, subscriptions, endowments, revenues from dinners or tag days, and an occasional paying patient. Their image, incentives, and management style were those of a charity intent on doing good, not a well-run business with a product to sell. Annual budgets reflected a balance between the need for nursing services among the poor and generosity by the affluent. To stimulate charitable impulses, annual reports traditionally pictured caring nurses rescuing patients from disease and poverty. Innovations and experiments in community-based care often rallied additional support. When lady managers inevitably found themselves in what Mary Gardner called the "rather embarrassing position of a man who had suddenly set up a large establishment without any particular increase in income," they counted on their friends to open their pocketbooks. To these ladies, a little overspending was of lesser concern than a surplus, which seemed to represent work undone.[1]

Initially, VNAs with small budgets ran few risks, but in about 1912, as budgets began exceeding $50,000 annually, problems began to arise. Fundraising quickly became a far more serious challenge, and "appalling deficits" were common. Managers in some cases described "an almost desperate sense of being overwhelmed by sudden growth and extension of this kind of enterprise which they had helped into being . . . without a correspondingly rapid increase in sources of support."[2]

Lillian Wald's Experiment

As a result of Lillian Wald's groundbreaking work with the Metropolitan Life Insurance Company (MLI) in 1909, insurance payment for home nursing care would dramatically change the financial circumstances of many VNAs. Dr. Lee Frankel, founder that year of the MLI's Welfare Division, believed that a program of protection from the ordinary risks of industrial life could reduce poverty and dependence. Motivated by the amount of illness that company agents reported among policyholders, he created an agenda that linked industrial insurance, social welfare, and good health. The MLI's insurance agents were chosen as the army to serve Frankel's campaign for health.[3]

Creation of Frankel's Welfare Division was certainly influenced by the Armstrong Commission Hearings of 1905–1906. Extensive newspaper coverage described in detail the insurance industry's ruthless competition, mismanagement of money, and flagrant disregard for policyholders. Following indictments for grand larceny, forgery, and perjury, the financial dealings of insurance companies came under more careful regulation. Compared to the big three—New York Life, Mutual, and Equitable—the MLI emerged essentially unscathed. Even so, legislative and social pressures for industry reform mandated that all insurers move beyond a purely entrepreneurial agenda. Speaking of Metropolitan's Welfare Division, Frankel later recalled that it became "more and more evident daily, both in our legislative enactments and in the thought of the general public, that life insurance [was] not a business, and that it [could] no longer be conducted along purely business lines." Now it was a social institution.[4]

Wald was intrigued by Frankel's ideas and thought home nursing care would be a cost-effective social investment for the MLI. Frankel invited Wald to meet with MLI directors and to bring a proposal and "documentary evidence of the nurses' practical use." She provided impressive statistics comparing pneumonia patients receiving nursing care at home with those in hospitals. Wald proposed that Metropolitan "exploit the good will, reputation, as well as ability of nurses organized in reputable outstanding agencies, to provide nursing services to their policyholders." For a modest fee per policy, the MLI could, she argued, reduce the number of death benefits paid. Such an arrangement would also mean that, without additional fundraising, visiting nurses could extend their services to more working-class people. Arguing that the mutual advantages were readily apparent, Wald recommended that the MLI test her plan with a brief experiment.[5]

The MLI agreed to Wald's plan to compare the outcomes for sick policyholders receiving visits from Henry Street Settlement nurses in one section of the city with those not receiving visits in a comparable section of the city. The experiment began on 1 June 1909, with Ada Beazley's visit to an MLI policyholder on Hudson Street. Ella Crandall, who would later become the executive director of the National Organization for Public Health Nursing (NOPHN), oversaw the experiment. Within three months, the results were convincing enough for the "wise directors" to extend coverage to policyholders throughout the city.[6]

The MLI's field agents functioned as "door openers," notifying the Henry Street Settlement of sick policyholders. A nurse then visited the family, provided care, and sent a report to Metropolitan stating the diagnosis, number of visits, and care given. On the basis of these records, the MLI paid Henry Street 50 cents per visit, a price set by Henry Street as a best guess, "a nice round figure—for at that time no one knew how to figure actual cost."[7]

The experiment initiated the first of 107,500,000 home visits to 20,150,000 MLI policyholders between 1909 and 1952, at a cost of $115,700,000.[8] By 1911, Mother Met, as the nurses affectionately referred to the MLI, had extended its nursing services across the country. In 1916, the services of visiting nurses were available to 90 percent of the company's 10.5 million industrial policyholders living in two thousand U.S. and Canadian cities. That year, 221,566 people received 1,189,828 nursing visits at a cost of $612,935. Between 1911 and 1921, the MLI spent 6 percent of its administrative budget on the Welfare Division; of this, 40 percent went for the nursing service ($306,000 in 1911 and $1,400,000 in 1921). The general policy of the company was to contract with existing VNAs when possible. Where no agency existed, individual nurses were employed on a salary or paid per visit.[9]

The principal objective of the MLI's contracts with visiting nurse services was the restoration to health and work of any policyholders who were ill. The economic incentives were obvious. By increasing the life span of policyholders, fewer death claims were filed, premiums were lowered, and this, in turn, attracted more policyholders.

Who's in Charge?

The Metropolitan Life Insurance Company operated essentially as a mutual aid association to safeguard its members against "certain contingencies of life." Funds received from policyholders and investments were used first to pay claims, second to create a reserve fund for future claims, and third to pay overhead expenses. Whatever "savings" remained provided the resources for the MLI's "welfare work" and reduction of premiums. Metropolitan's financial support of visiting nursing was hardly just philanthropic, and even at a cost of 5 cents per policy, restrictions applied on the amount of service provided. Visiting nurses were an extra privilege, not simply a routine part of the insurance contract between company and policyholder.[10]

Although Henry Street Settlement's limited experiment was a success with MLI policyholders, public acceptance and financial feasibility could be evaluated only with a larger-scale offering of services. Outcomes remained ambiguous in 1911, although it was quickly apparent that success meant careful management of cost, demand, productivity, and mortality.[11]

In order to extend the visiting nurse service to the working class, the cost of a visit had to be maintained at about 50 cents. The MLI assumed that visiting nurses worked eight hours a day, six days a week, and averaged eight to ten visits a day, or two hundred a month. Nurses were paid an average salary of $75 per month. From this information, the MLI calculated a cost per visit (salary divided by the number of visits) of 35 cents. Management and travel costs produced an actual cost of 45 to 50 cents per visit. Nursing service was used by one to three patients per thousand policies, with each patient allocated an average of five to six visits. Thus, one nurse was required for every twenty thousand policyholders.[12]

The MLI's second parameter of financial feasibility was a variation of input and output calculations based on "lethal rates," number of deaths per hundred cases of illness. These percentage rates were used to identify the distinct value of nursing care in the treatment of specific diseases. While claiming it was "adverse to drawing any definite conclusions" from these calculations, the MLI quickly discovered that visiting nurses made more than five visits to most patients and for some illnesses provided care that demonstrated little practical financial return.[13]

On the basis of this information, in 1911 Frankel asked the VNA boards in Chicago and Boston to hire practical nurses to care for patients with less complex, chronic conditions. With these less expensive "nurses," Frankel hoped the costs of care would be reduced. Suddenly VNAs could no longer ignore the power of Mother Met.[14]

Nursing leaders discussed the matter of practical nurses for chronically ill patients privately with Frankel and publicly in the nursing journals.[15] In an article outlining the risks to patients, Edna Foley, superintendent-elect of

Chicago's VNA, asked rhetorically whether Frankel's request meant that VNAs were "intended for the sick or to act as investigating agents and supervisors for insurance interest." Had standards been "so lowered that the number of visits and the amounts of instruction given count, while the actual work of our hands has become so unimportant that it can with safety and expedition be handed over to so-called practical nurses, whose practice is on par with the scanty remuneration they receive?" Foley concluded that "the poor are at the mercy of too many half-trained and counterfeit workers as it is, and it behooves visiting nurse associations in good standing to maintain the integrity of our calling by offering their best alike to the acute and chronic sick."[16]

Despite the strong stand against Metropolitan's proposal, many nurses did in fact recognize that chronically ill patients did not require expensive expert care. But the struggle to convince the public that professional expertise was important compelled nurses to resist all potential threats to practice or employment. In the end, Frankel dropped his plan, instead deciding to limit the number of reimbursable visits to chronically ill MLI policyholders. By 1915, chronically ill patients received twenty thousand fewer visits than when the program was initiated in 1913.[17]

Experience with the dictates and demands of the MLI further heightened the need for visiting nurses and their lady managers to organize themselves. According to Isabel Lowman, organization was now thought necessary if "the treasures of their tradition" were to "be preserved intact." At the national nursing convention in Chicago in June 1912, the NOPHN was created and Lillian Wald was elected president.[18]

The NOPHN formed none too soon, for Frankel's next target was the "charity image" of visiting nursing. In December 1912, he asked three of the larger VNAs to separate their staff into two groups, one for charity cases and another for MLI policyholders. The nurses doing "Metropolitan" work wore a "brevet" (a badge to distinguish them from charity workers) and had their own supervisor who managed MLI patients in a businesslike manner. Creating at least the illusion of distinctly different nursing services shielded policyholders from the appearance of accepting charity or minimized their reluctance to be cared for by a nurse from a charity service.[19]

Hearing of Frankel's latest scheme, Ella Crandall, executive secretary of the newly formed NOPHN executive committee, prepared a response for the *Visiting Nurse Quarterly* in which she concluded that VNAs could not share the business world's motives and would not become propagandists for commercial concerns. VNAs preferred to "withdraw from the present connection with the company [rather] than to imperil the spiritual and social implications that are involved in their present relationship to the community."[20]

Frankel reviewed the article in advance and was offered "the privilege of answering in the same issue." Frankel was indignant and resentful about the proposed article, but Crandall was adamant, declaring that "if the new organization [NOPHN] has any purpose at all, it is to warn its constituents in advance instead of later." The disadvantages of Frankel's plan were, in her opinion, obvious. Although the lady managers considered insurance patients a natural extension of their work and were willing to improve their record keeping and to admit that their nurses could make more visits per day, they thought Frankel's latest plan went too far. Frankel agreed to postpone his experiment and Crandall's article was never published.[21] Nevertheless, Frankel continued to believe that the MLI's service was seriously hampered by the stigma of charity and countered that image in a variety of ways, including publishing a brochure for policyholders titled "Nursing Service Is Not a Charity."[22]

Crandall wrote to the NOPHN executive committee to remind its members that, although she had negotiated an amicable adjustment, Frankel's experiment was postponed, not abandoned. She cautioned all to come to the national nursing convention in June prepared for more negotiations. Mary Beard later remembered these early confrontations with amusement. The nurses actually feared Frankel, she recalled, and it seemed impossible to interpret "our professional standards to him or his profession's needs to us."[23]

Fortunately, the June 1913 convention had a catalytic effect on the relationship between Frankel and the NOPHN. By invitation he publicly presented his concerns in the paper "Visiting Nursing from a Business Organization's Standpoint," which was published the following month in *Public Health Nursing Quarterly.* Frankel warned that "if visiting nursing services are to receive the development they should, and are to reach the large mass of the population, less must be constantly laid upon the charitable or philanthropic side." Visiting nursing should be conducted as a business proposition and should support itself largely through patients' payments, paid per visit or annually by employers, fraternal orders, benevolent associations, labor unions, insurance companies, municipalities, or relief agencies.[24] By the end of the convention, the NOPHN's executive committee and Frankel realized they actually shared a common goal. What remained unclear was how it would be achieved.[25]

With the MLI providing 20 to 30 percent of their annual budgets, most VNAs could hardly ignore Frankel's opinions. When word of the MLI's plans to organize its own visiting nurse service circulated, Frankel's concerns took on a new seriousness and were regularly quoted in annual reports. As Mary Gardner suggested, whether or not one agreed with Frankel, his views remained a "burning idea."[26]

Metropolitan's Policyholders

The recipients of these visiting nurse services were the weekly-premium-paying policyholders of the MLI's Industrial Department. Industrial policies were designed to do for the industrial classes what ordinary insurance did for those in better circumstances—namely, to provide funds for the family during the critical period immediately following the death of the policyholder. Unlike ordinary insurance, these policies could be purchased for women and children, as well as men, without a medical exam. The simplicity of the contracts and reasonable cost made these policies very attractive to low-income workers. Weekly payments ranged from 3 to 15 cents, for policies with a face value sufficient to cover burial expenses and provide a small amount of cash for survivors' security ($50 to $250). Payments were collected weekly by MLI agents at the policyholders' homes.[27]

The number of industrial policyholders increased from eight million in 1911 to seventeen million by 1935. Between 1911 and 1916, 87.5 percent were white and 12.5 percent black. A little over one-third of policyholders were between one and fifteen years of age, and over half were female. Despite the dominance of women, occupations of policyholders were reported only for men over fifteen years of age. Most occupations were represented, reflecting policyholders' largely (95%) urban residence.[28]

Like most insurance companies, the MLI had a tenuous relationship with the black community, as reflected in its recurrent acceptance and rejection of "Negro risk." Because of their higher mortality, blacks were often sold policies with severely limited benefits or charged higher premiums, or both. Beginning in 1907, this so-called extra mortality prompted a new plan to base the cash values of black policies on special tables and dividends on actual mortality experience. Blacks were also subject to "careful selection" and, unlike whites, were given a full medical exam. Despite the MLI's precautions, death rates remained substantially higher for blacks than for whites. And even with these efforts to charge black policyholders a higher premium than whites, business among blacks was described as not profitable. As late as 1928, the MLI remained conflicted about its commitment to writing these policies.[29]

Of particular concern were the prevalence, acuteness, and seriousness of sickness among rapidly growing urban black populations. Examining the "health prospects of the American Negro" in 1927, MLI statistician Louis Dublin saw "the Negro playing an increasingly important and worthy part in Americans' affairs, and regard[ed] his achievements as the greatest experiment in racial adjustment ever undertaken by man and as a most encouraging and gratifying episode in our national life." He expected the nursing service to play an important part in this agenda, benefiting the community and helping the MLI balance its financial responsibilities and welfare responsibilities to the black community.[30]

"An Era of Health Promotion"

Initially, the MLI wanted to reach as many policyholders as possible. Between 1909 and 1915, the number of policyholders visited increased by 202,659. Despite this growth, in 1915 only 14.46 of every thousand policies in force involved the services of visiting nurses. Although a significant increase over the 1910 rate of 3.08 per thousand, this low ratio raised the question of whether the nursing service was reaching the policyholders most in need. The MLI conducted studies of death claims in several cities and concluded that the ratio was in fact a "fairly true gauge of the service needed." The most extensive analysis of death claims was conducted in Baltimore, where only 8.9 percent of 2,968 death claims were for patients who had received nursing care. For half the total death claims, however, this lack of care could have been because the people died in institutions or as the result of suicide, homicide, accidents, or sudden illnesses, or lived in areas without visiting nurses. For the remaining policyholders who died, most did not qualify for nursing visits because they suffered from chronic illness.[31]

Because black people represented over 12 percent of policyholders, the MLI particularly wanted VNAs to reach out to the black community, providing care and reducing mortality rates. Dublin reminded the nurses in 1921 that "Negroes are often less able to provide individual [private-duty] nursing care and have usually higher sickness rates than white people." VNAs must therefore care for an "even larger proportion of cases for colored people than is indicated by their distribution in the population." Although on average black policyholders received proportionately more care than whites, the MLI wanted VNAs to keep "figures by color" so that it could document progress and anticipate changes in risk.[32]

National Standards with a Few Local Degrees of Freedom

As mortality rates among policyholders declined compared with those in the general population, the MLI became convinced that its policyholders were being reached.[33] The company proceeded to implement the rest of its administrative agenda launched at the national nursing convention in 1915. Armed with several years' worth of statistics, Lee Frankel presented and later published his paper "Standards in Visiting Nurse Work." Based on a study of twelve VNAs considered typical of the best nursing services in the country, Frankel advocated the evaluation of VNAs' performance against a national standard. He suggested that accounting for the work done was a first step toward determining whether the results justified continuation of this work. Ella Crandall, who moderated Frankel's session, concluded that the "easy going casual, comfortable way in which we have done the work on a purely philanthropic basis" needed to be replaced with a "sound business-like program and method of procedure."[34]

Publicly, the MLI claimed that the twelve VNAs described in Frankel's paper recognized the fundamental principles of visiting nursing and were under competent management, but private evaluations were much more critical. The MLI's analysis of their conformity to standards ranked the performance of several respected agencies as less than stellar. For example, while Chicago's conformity to MLI methods was 98 percent and Henry Street's was 92 percent, Boston's and Philadelphia's "conservative administration, and tendency to continue the old regime" were rated at 80 percent. Brooklyn's "evident lack of nursing standards and unsystematized administrative methods" resulted in a rating of 70 percent.[35]

The MLI's 1918 study confirmed earlier findings that home care practices varied dramatically from one agency to another. In New York, for example, pneumonia patients were visited daily for thirteen days, 12 percent were referred to hospitals, and 79 percent recovered. In Baltimore, however, pneumonia patients were visited every day and a half and received only 6.5 visits, 45 percent went to the hospital, and 19 percent recovered. Clearly, local cus-

toms not only determined practice patterns but also impeded the establishment of common standards and nationwide agendas.[36]

Although continuing its quest for a standard that all could emulate, the
MLI attributed the lack of uniformity to the degree of development in each
city. The company acknowledged the need to examine the "influence of
strictly local circumstances."

> The experience shown for a public health nursing society is in a very large mea-
> sure determined by such important *local* factors as the extent of hospital facil-
> ities in a given community, the degree to which physicians cooperate with the
> association, the attitude of the population served toward the work of the as-
> sociation, the density of the population, the nature of the industries which
> provide the means of livelihood for the community reached, and the funding
> conditions for philanthropic work in the local area.[37] (emphasis added)

Case Management: Cost Containment
by Administrative Controls

Although Mother Met sought standardization of practice, the company also
understood the need to pursue a reasonably elastic administrative program
to meet local conditions. The MLI's next strategy was to use "intensified
administrative measures" to eliminate "useless cases." Through a system of
constant correspondence, close supervision, and regular visits of inspection,
case management became the major focus of the MLI's program for life extension. The evolving components of "practical" case management eventually included (1) proper case selection emphasizing acute illness, maternity
care, and the acute phase of chronic disease, plus the vigorous exclusion of
chronic illness from covered care; (2) proper discharge of patients when
nursing was no longer required; and (3) early reporting of cases for care.[38]

The MLI used time between visits to analyze the degree of success in case
selection. A considerable period between visits indicated that nurses were
not getting the acute type of cases in which actual nursing accomplished the
greatest results. In 1914, 2.2 days elapsed between visits; in 1916 this declined
to 1.6 days. For Metropolitan this indicated that the more suitable acute cases
were being reached more frequently during the critical period of illness.[39]

Efforts to ensure proper case management between 1914 and 1916 decreased the average duration of care from eighteen to thirteen days. By 1915,
visits per case declined to 5.3, compared with the 9.28 visits per case paid for
in 1910, representing substantial savings.[40] The MLI also computed "an in-

dex figure" for success in securing "proper" types of cases. By 1916, 30 percent of MLI cases were acute, 9 percent chronic, and 22 percent maternity. Between 1914 and 1916, changes in these indexes documented a slight increase in the number of acute cases, a substantial decline in chronic cases, and an increase in maternity cases.[41] Most gratifying to the MLI was the gradual reduction in the amount, duration, and intensity of service to the chronically ill. By 1916, the company was paying for approximately 25,490 fewer visits to chronically ill policyholders than in 1914. The most striking decreases in average costs per case were achieved for chronic conditions. Despite claims of success, the average chronic case still cost $1.50 more than the average case ($4.12).[42]

The MLI remained very attentive to the importance of the early reporting of cases. Those reported soon after onset of the patient's illness produced maximum nursing success, or so it was believed; thus, referral source was selected as a proxy for timeliness. Even with local variations, the MLI concluded that referrals from families or physicians resulted in the greatest success. Cases reported by the agent were not conducive to good service, since waiting for agents' visits usually meant valuable time lost in getting care. In 1923, the MLI found 63 percent of cases were reported by agents, 22 percent by families, and 4.4 percent by physicians. By 1931, referral sources had changed dramatically: 58 percent from families, 32 percent from agents, and 5 percent from physicians.[43]

Cases proving to be a waste of time (27%) included those in which the nurse found no illness, no patient, no policyholder, no physician, a recovered patient, or herself unwanted. The MLI denied reimbursement in these cases—an expensive mistake for the VNAs.[44]

Following passage of the Sheppard Towner Act in 1923, with its infusion of funds into local maternal and infant programs, the MLI reconsidered the amount of maternity care it needed to provide. Under this new set of circumstances, the company declared prenatal work the responsibility of official health departments. It examined with a new eye the 30 to 70 percent of visits to maternity patients and concluded that maternity care for policyholders exceeded requirements. By deducing that maternity care by VNAs resulted in less care for patients with acute conditions, the MLI decided that the number of maternity visits could be reduced without risk to life. Once again, Metropolitan set out to adjust its policies.[45]

Along with carefully prescribed administrative procedures, the MLI also increasingly relied on agents and medical examiners to serve as "guardians at the gateway of the company." As a rule, medical exams were not required for policy applicants, but all potential policyholders were expected to certify a claim of sound health. Applicants were excluded if "health, occupation, or moral condition" was likely to lead to early death. Of primary concern were policyholders of short duration—those who died from "selective causes" within one year of purchasing a policy. These cases were predominantly patients with chronic conditions, but a few with onset of illness within six months to a year were especially costly mistakes. In 1914, for example, death claims paid to 3,627 policyholders of less than one year's duration cost the company $420,513. By the 1920s, careful selection of black applicants was of such great concern that some insurance companies hired black agents to screen black applicants. Experience had demonstrated that the prejudices of white agents prevented their being able to discern good "moral risk," whereas black agents' knowledge of "the classes and groups" in the black community made them more sophisticated in "selection of colored risk." From the company's perspective, better selection resulted in a "better class" of policyholders and ultimately improved mortality rates.[46]

A Search for Greater Harmony

By the 1920s, Frankel claimed that nursing associations and nurses everywhere understood the goals of the MLI. He believed that scientific case management meant timely referral of acutely ill patients, a few closely spaced visits, and discharge within two weeks.[47] The nurses may have understood the MLI's case management agenda, but their interest in it was less certain. Predictably, Dublin described "nursing work" as still in an unsettled state. While hoping for greater harmony, he found instead that associations across the country had a great variety of procedures, aims, and even types of organizations and personnel.[48] Reviewing the performance of the top fifteen VNAs in 1923, the MLI reported "tremendous variations in the cities studied with respect to the work we are doing . . . especially in terms of professional judgment."[49]

According to Metropolitan, good professional judgment occurred when a case involved the requisite number of visits, suitable nursing treatment, and quick disposition in the best interests of patient and family. In 1923, the company rated professional judgment as good in only 58 percent of cases examined, suggesting a continuing "diversified waste of effort." Top performance ratings went to Philadelphia (99.7%), Indianapolis (99.5%), Boston

(98%), and Brooklyn (97%). At the bottom were Rochester (27%), York, Pennsylvania (32%), Oklahoma (43%), and New York City (51%).[50]

Despite its patient-centered rhetoric, the MLI's major concern remained cost containment. The company's statisticians knew very well that "the number of visits [was] the measure of the service, and if control [was] to be exercised, it [was] through increase or curtailment of the numbers of visits, with due regard to the prevalence of respective diseases and their relative hazard to life." In addition to relentless efforts to promote its concept of case management, the MLI also carefully analyzed visit patterns. During cursory reviews, clerks flagged flagrant variations in standards: more than twenty visits, patients acutely ill after four weeks, more than six visits to the chronically ill, and more than twelve visits for hypodermic injections for the acutely ill. Questionable cases were referred to a Professional Reviewing Unit made up of nurses.[51]

This unit searched for evidence of "excessive" visits or inappropriate requests for reimbursement. Certain practice patterns were clearly scrutinized: (1) exorbitant numbers of cases or visits, (2) excessive visits for certain diagnoses (i.e., a high percentage of acute cases), (3) excessive numbers of visits to chronically ill patients claiming hopeful prognoses, and (4) too many prenatal cases requiring extra visits because of patients' blurred vision. Beyond evaluating "professional judgment in case management," the Professional Reviewing Unit also identified cases "not acceptable for nursing care," such as those with no physician, midwife cases, convalescing patients, nurse not wanted, patient transferred to hospital, or no policy. Payments were denied for inappropriate visits. Denials varied among agencies but averaged an amazingly high 24 percent of cases reviewed between 1923 and 1929.[52]

Frankel's final role as nursing's unsolicited business advisor was to sponsor an impartial study of the work of VNAs to examine both quality and quantity of care. Information about the most "simple facts" in the majority of agencies was sparse and findings were limited, with conclusions more qualitative than quantitative. The picture of visiting nursing that emerged revealed a muddled state of affairs. Extreme variability in services made comparisons of quality and cost of care nearly impossible. The study was unable to determine what constituted the usual nursing workload or the frequency of care in various types of cases. Finally, costs per visit could not be compared owing to variations in accounting methods.[53]

The MLI remained concerned about missed opportunities, wasted effort,

TABLE 1. Philadelphia Cases, 1923

	Percentage of Total		Average Visits per Case	
	Metropolitan	General Service	Metropolitan	General Service
Acute diseases	52.7	22.8	5.6	13.2
Chronic conditions	12.1	19.7	8.0	20.7
Maternity	32.0	53.8	8.1	11.5
Other	3.2	3.8	8.7	20.1

Source: Metropolitan Life Insurance Company.

and excessive variability among the VNAs providing services to policyholders. By 1923, however, changing practice patterns suggested that case management was producing at least some of the desired results. Statistics for Philadelphia, the MLI's chosen exemplar for adequate and appropriate care, demonstrated increasing attention to more acute cases, declines in chronic cases, and a more conservative approach to maternity care. Especially in terms of visits per case, the difference between MLI cases and general service cases was striking (Table 1). In 1923, Metropolitan paid for care in 30,854 cases in Philadelphia, with 203,636 visits. With general-service patients receiving twice as many visits as MLI policyholders, at a cost of 80 cents per visit, case management saved the company $162,909.[54]

Nursing "Propaganda"

The MLI Welfare Division's "health work" also included pamphlets (Figure 14), community work, exhibits, clean-up campaigns, and movies. The company's health campaign reached twenty-three million people, with 420 million pieces of literature distributed. By 1930, MLI nursing services were located in four thousand communities. Nurses had provided care to nearly eight million policyholders and made over forty-two million visits, at a total cost of nearly $34 million. The service grew each year. Cost per visit also steadily increased from 44 cents to $1.02. Following a first-year all-time high of 9.28 visits per case, by 1930 visits hovered around an acceptable five per case.

The MLI claimed a greater reduction in mortality among its policyholders than in the population at large, as well as "the greater reduction in those conditions which are more predominately represented in the records of the visiting nurse service . . . the best reduction appear[ing] where the greatest

FIGURE 14. Metropolitan Life Insurance Company Pamphlet. *MetLife Archives.*

nursing effort has been expended." Although agents were not supposed to use the visiting nurse service as a "sales argument," it was a "valued privilege" and as such a wonderful form of propaganda for the MLI. The establishment of nursing services by John Hancock, West Coast, Travelers, and Aetna Life insurance companies in the mid-1920s confirmed the growing popularity of this privilege.[55]

Lee Frankel would at last declare that "stabilization" was achieved. According to his standards of high efficiency and low cost, Metropolitan's nursing service was operating as an efficient business. Maternity visits had declined, average visits per case had decreased to 4.9, the cost per case was 99 cents, and the percentage of good records had increased to 94 percent. Most gratifying was the stability of these accomplishments after 1923. Despite private anxieties about the nursing service, Metropolitan showed no hesitation in publicly declaring its success in reducing mortality and saving dollars. Mortality savings, by the company's calculations, were more than sufficient compensation.[56] Typical of MLI claims was the following statement in the March 1929 Statistical Bulletin:

> By the close of 1928, the death rate of Industrial policyholders had declined to 8.6 per thousand, or 31.2% less than the rate which prevailed seventeen years before. The rate of decline was twice as great as in the general population, and this was true not only of the total mortality but for most important diseases and conditions. The decline in mortality of policyholders, over and above that in the general population, was clearly a measure of the effectiveness of the efforts of the company in health education, the nursing of the sick, and the other measures which had been instituted.[57]

"Limitations of the Claims"

Of the known criticisms of claims that the nursing service contributed to the declining mortality of policyholders, the most telling was published in 1933 by Maurice Taylor, director of the Jewish Family Welfare Association of Boston. Taylor's particular interest was the social cost of industrial insurance. While wanting to avoid discrediting the contributions of the MLI's nursing service, being aware of its advertising worth to the company and accepting the assumed value of nursing care, Taylor viewed with uncertainty the net health gains claimed by MLI statisticians.[58]

Taylor gave three compelling reasons for his conclusion that comparisons between MLI policyholders and the general population were not legitimate.

First, beyond the obvious universal reduction in mortality rates and overall improved standards of living of the general population, the insured class was much younger and more urban than the general population and therefore would be expected to have lower mortality. The MLI's efforts at "selection," if at all successful, ensured that policyholders were healthy enough to work and likely to remain in reasonable health. The second issue was the MLI's high "lapse rate." At its lowest in 1920, the company's policy turnover was 28 percent; by 1928 it had reached 41 percent. Such turnover added considerable uncertainty to claims of health benefits gained, since it was unclear what proportion of the original population remained to be counted. If lapsed policies occurred among the poorest policyholders, then the remaining policyholders were financially better off and healthier. In any case, high turnover made it very difficult to calculate the policyholder population at risk of dying.

The third problem was disentangling the effects of nursing intervention from other factors affecting mortality rates. For a large number of diseases, mortality rates were declining many years before Metropolitan's nursing service was introduced. Closely related to this fact was the low proportion of policyholders actually cared for by the nursing service. Despite the rapid expansion of the service, at best only thirty-five of every thousand policyholders received nursing care each year. Even assuming that in these cases death rates were reduced to zero, the effect on the total death rate would still remain negligible. Finally, in addition to these three arguments, comparisons of mortality rates of MLI policyholders with those of an equivalent group of policyholders insured by Prudential Insurance Company, which did not provide nursing services, found lower mortality in the latter group. In 1928, the MLI's age-adjusted death rate was 883.6 per hundred thousand; Prudential's was 869.

Taylor concluded that his arguments did not detract from the value of the MLI's health and welfare work, but did suggest the company's results were not measurable in the aggregate. In other words, "They [the benefits of a nursing service] may nevertheless be present. The money is well spent and should be continued, but caution is necessary in interpreting the results." Taylor's critique was unarguable, and even a statistician of Dublin's fame and expertise could make no rebuttal. It remains a mystery today why Frankel and Dublin chose to ignore the fallacies of their claims that the visiting nurses had dramatically reduced the mortality rates of policyholders. Despite claims to the contrary, the MLI's investment in nursing was clearly motivated by more than a goal of conserving life.[59]

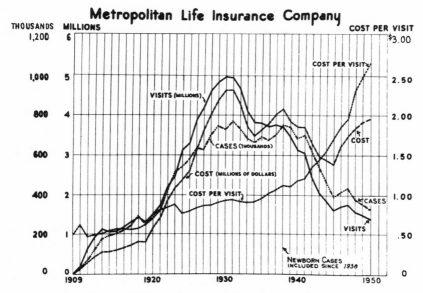

FIGURE 15. Nursing Service: Trends in Visits, Cases, and Costs, 1909–1960. *Metropolitan Life Insurance Company, MetLife Archives.*

The Ending

The Metropolitan Life Insurance Company's nursing service continued to grow until the Depression, peaking in 1931 with 770,000 policyholders receiving care (Figure 15). Lee Frankel died in the same year, and under new leadership the MLI's attitude toward the nursing service changed.[60] With the Depression came canceled policies, declining service volume, and rising expenditures. From the perspective of the insurance company, this combination made visiting nurses seem a less economically viable method of preventing death or attracting customers. These compelling realities notwithstanding, the demise of Metropolitan's nursing service was predictable. MLI policyholders lived longer, their mortality rates had decreased almost 50 percent, and they were increasingly affluent and able to purchase necessary care. Acute communicable diseases were no longer the major causes of death, and care of acutely ill and maternity patients was becoming the responsibility of hospitals. With the growing centrality of the hospital, fewer policyholders needed or used the visiting nurse service; those seeking the services of a nurse were most often the chronically ill elderly. From an insurance perspective, nursing intervention in these cases rarely reduced the death benefits paid by the company and was therefore considered a poor investment. By the 1940s, the MLI was spending $4 million annually to care for 1 percent of their policyholders whose needs no longer matched the original mission of the nursing service.[61]

Despite the dramatic decline in demand for the nursing service, in 1947 the MLI's Advertising Research Bureau still found "much evidence of a large potential demand for the type of home nursing service offered by the VNA and the insurance companies." It concluded that the almost universal appreciation of those who used the services was reason enough to assume that others, when sick at home, would take advantage of the service if they only knew how to get it and how much it benefited others. What was needed was more publicity![62]

The decision to discontinue the service was made in July 1950. Even Louis Dublin conceded he could no longer find data to show that nursing care lowered mortality rates or improved the company's sales or image. The MLI "bade adieu" and declared its mission accomplished, but generously gave the VNAs two years to prepare for a 10 to 20 percent decline in income. In his letter to the field force announcing the end of the nursing service, MLI President Leroy Lincoln wrote, "Our withdrawal from this activity can only be viewed as in harmony with today's changed conditions." On 1 January 1953, it all ended.[63]

Once again, the vexing problem of chronic illness had tipped the delicate balance between home care's costs and benefits. Providing care for those who failed to recover quickly was, from an insurance perspective, a poor investment. Even after forty years of experimentation, determining what kind of care, and how much, should be provided proved too complex and risky. Despite the MLI's prudent system of case management, care at home remained an often "uncontrollable" family matter, and nurses always seemed to place their patients' care requirements above the need for cost containment. As Leroy Lincoln claimed, the MLI's decision was more a response to changed conditions than a disenchantment with home care.

Part IV

Reinventing Home Care
in the Mid-Twentieth Century

CHAPTER 8

"An Unchanging Purpose in a Changing World"

$\mathcal{B}$y the 1920s, hospitals were recast from a place of last resort for the urban poor into a medical center for everyone. For many families, the hospital offered an attractive alternative to family caregiving at home. Visiting nurse associations (VNAs), having recovered from the challenges of World War I and the influenza epidemic of 1918, enjoyed a few years of unprecedented standing with the public. Mary Beard described this as a time when VNAs "grew rapidly and developed irregularly."[1] But postwar expansion was followed by financial crisis, self-analysis, and a revisiting of the local politics of public/private responsibilities for the financing of home-based care.[2] The work of the visiting nurse now had little in common with its earliest history. "Dangerously ill" patients were encountered infrequently, communicable diseases were rare, and chronic illness was consuming as much nursing time as caring for the acutely ill.

A Time of Readjustment

These changes affected the VNAs' ability to attract and retain an adequate staff. Within a few years, many VNAs found that demand for nursing care exceeded staff size, such that agencies had to refuse cases. The adequacy of salaries was questioned, and salary studies, raises, larger staffs, and bigger budgets followed. In Philadelphia, for example, the nurses were especially concerned about low salary, claiming they were paid a subsistence wage. In March 1925, Katharine Tucker told the board of the Visiting Nurse Society of Philadelphia (VNSP) that the salary issue was simply a question of policy and nurses must be paid more, even if the number of staff and therefore the volume of work had to be reduced. Not surprisingly, the board voted to increase salaries. Tucker claimed that Philadelphia's nurses were paid less than those in any other large nursing organization (except those in Baltimore and Washington, D.C.), but in fact the VNSP paid the twelfth highest (of 398 VNAs) salary. Tucker believed the staff needed to increase by ten to twenty nurses. In one week in 1924, the society had to turn away 125 cases and, since priority was given to the acutely ill, it usually refused healthy maternity patients and the chronically ill. In Boston, Mary Beard used a similar approach to raise the nurses' salary. Had the board failed to raise the money, she planned to decrease the number of staff by enough to pay those remaining a "living wage."[3]

At the same time, many VNA boards had increasing difficulty raising the money needed for salaries and hiring the numbers of staff required by the growing demand for services. It was a precarious situation, given that even the most aggressive VNAs obtained only 30 percent of their budget from patients' fees (mostly insurance payments), with the remainder coming from investments and contributions.[4] For example, Henry Street Settlement's deficit of $46,872 raised many questions about overhead, cost, and production. The settlement decided it had to find a way to cut overhead without destroying "the real efficiency of the service." The decision was to reduce the staff by ten, increase the nurses' "production" (visits/day), and reduce the cost per visit by extending their daily hours.[5] The growing uncertainty about funding an ever-increasing budget thus loomed large for most boards. Such circumstances usually discouraged any attempts to undertake new programs. Innovation was replaced by pragmatic concerns about obtaining paying patients, efficiency in operations, and contributions.[6]

At the June 1924 meeting of the National Organization for Public Health Nursing (NOPHN), all local worries were temporarily overshadowed by the impending financial crisis of Boston's Community Health Association (CHA). Although dramatic, it was an all too familiar situation for nursing leaders who attended the meeting. It was also particularly ironic because most observers considered the CHA, under the guidance of Mary Beard, the ideal visiting nurse association. Boston's CHA was organized in October 1922 through the amalgamation of the Instructive District Nursing Association and the Baby Hygiene Association, to create what was described as a family health service. The CHA offered the citizens of Boston a "generalized" neighborhood nursing program, maintained high standards of care, and experienced a period of remarkable growth. While the average association provided care to twenty-one of every thousand residents, the CHA cared for sixty-eight of every thousand Bostonians. Unlike any other association, the CHA could plausibly assert its intention of extending the services of the visiting nurse to the whole community.[7]

To expand the CHA's work, seventy-five nurses were added to the staff after the war, and salaries were raised on three separate occasions, with a budget increase from $150,000 in 1919 to $357,000 in 1923. Operating in accordance with time-honored methods, Mary Beard presented her new plans and larger budgets to the CHA board each year, and the board in turn looked for the necessary support. As raising money became more difficult, the board sought the advice of some businessmen and decided to hire an "ad man," Mr. Morgan, to help them. The board was quickly informed that funds could no longer be raised in the ladylike way to which it was accustomed—tag sales, annual subscriptions, and charity balls. A vigorous publicity campaign was "a man's job," said Morgan. He called for advertising in newspapers; posters in train stations, Copley Square, and Boston Common; and active solicitation by "girls" in hotels, department stores, and places of public congregation.[8]

The CHA board saw Morgan's plans as too aggressive, and his activities were limited to writing advertisements. Continuing the search for an acceptable method of fundraising, a board member was sent to Philadelphia, Providence, and New York to study the procedures of their VNAs. At the conclusion of these visits, the most successful approach was deemed an educational campaign conducted by (1) a businessmen's committee, (2) local neighborhood committees, and (3) societies of young women. This conservative approach proved acceptable to the board and was immediately implemented. For example, a male executive committee of nine members was appointed in 1920.[9]

The new strategies for fundraising apparently worked until the winter of 1923, when it became obvious that the CHA would end the year with a deficit of $26,000. The ladies turned to the male board members for help, but found them unwilling to put in the additional time on fundraising. The men thought a general organizational man was needed to supervise the finances and office management and to increase the organization's earning power.[10]

Even Mary Beard conceded that the question of limiting the work of the CHA was bound to arise sooner or later. The demand for service was finally reaching the point at which the staff could no longer increase with it, owing to lack of money or lack of properly trained nurses. The board's idea of limiting the work was very difficult, reported Beard, because this would mean reductions: first, in health-promotion activities, second, in prevention of disease, and third, in the care of the sick.[11]

Six months later the inevitable happened. The spring 1924 appeal for funds was a great disappointment, raising $34,000 less than anticipated. Twelve nurses were dismissed in April, but expenses would have to be further curtailed unless the board raised an additional $14,000 per month.[12] This proved impossible. Mary Beard was asked to present a plan for "curtailment of the work" to the executive committee at its May meeting.

Despite Beard's frustration and disappointment, the cuts had to be made somewhere. In her opinion they should come from maternity, chronically ill, or child health work, not from services to the acutely ill or services that were growing and paying for themselves. The supervisors had already paid a great deal of attention to the chronically ill caseload and had substantially cut the numbers of visits. Because no one else would care for such chronically ill patients, it would be difficult to cut this service further. In the end, the supervisors decided to limit their work with preschool-age children and to discharge all prenatal patients who planned to deliver in the hospital.[13]

The board would not vote on a final decision to reduce the CHA's budget by $130,000 until the week after the June 1924 NOPHN convention, but the outcome seemed obvious. At the convention the topic was unavoidable. The immediate crisis was finally resolved in August, when Boston's health department took over all child health work, allowing the CHA to reduce its staff by fifty nurses. The association ended the year, no doubt much wiser, with a deficit of $57,881. The $130,000 reduction in expenses was no longer necessary.[14] While the CHA's rapid expansion and retrenchment were extreme in comparison with the experiences of most VNAs, the basic issues were similar.[15]

Not surprisingly, the focal point of the NOPHN convention, held in Detroit, was a paper presented by William Norton, Secretary of the Community Fund of Detroit, on "Meeting the Demand for Community Health Work," followed by a roundtable discussion.[16] The message was not new, but when coupled with the news from Boston its impact was spectacular.[17] While some assumed that Boston's lack of a community chest precipitated the CHA's financial crisis, the secretary of the Detroit Community Fund was describing to the convention a much larger problem that could not be cured by so simple a solution as cooperative fundraising. According to Mary Gardner, the message was clear. For VNAs, the age of discovery, aspiration, experiment, and belief without proof was over; it had been replaced by the more practical "period of the proof of the pudding."[18]

After much discussion, the convention members agreed on five points:

First, that in all probability they could not expect in the next ten years the same rate of financial increase they had experienced in the past ten.

Second, that methods accepted in private life should apply to organizations—namely, no debt, and no expenditure of principal except under the rarest of conditions.

Third, that stricter business methods should be applied to publicity and advertisement.

Fourth, that educational work, as well as bedside nursing, should be charged for and there should be a more systematic effort to make all work self-supporting.

Fifth, that there should be general readiness to turn over the work of private organizations to public administration.[19]

Having committed to a more conservative, businesslike course of action, all agreed that executives who found themselves unable to balance their budgets should feel no guilt, for it was simply a sign of the times. Not only was a slowing down inevitable, but some believed a certain "curtailment" should occur. This, declared Mary Gardner, might not prove an unmitigated evil, "since with it [would] come a stricter analysis and appraisal of the various types of work in light of actual results obtained."[20]

A Time of Self-Analysis

Adopting "the modern attitude" of scientific study, for the remainder of the decade visiting nurses became obsessed by self-analysis and appraisal. They examined appropriate educational preparation for nursing practice and output-to-cost ratios in terms of quality, time per visit, interval between visits, analysis of visit content, caseload, case mix, record keeping, average time expended by type of health problem, and even cost per minute. As data accumulated, nurse-investigators produced estimates of "nurse power" and attempted evaluations of progress. These early investigators assumed that their research techniques would eventually become sophisticated enough to allow comparisons among various methods of work and would ultimately produce practice norms or generalizable standards.[21]

While promoting a more "intelligent appreciation" of the value of self-analysis, these studies barely acknowledged Mary Gardner's simple plea to plan for the future. Neither the questions asked nor the conclusions reached deviated significantly from earlier investigations. All focused inward, on issues of productivity and financial survival, rather than outward toward end results and changing community needs. Common to virtually every study, report, or analysis of visiting nursing at this time was the celebration of nurses' "unchanging purpose in a changing world." Predictably, one critical non-nurse reviewer pronounced these analyses uninspirational, the conclusions questionable, and visiting nurses' place in health care unclear.[22]

Meeting in New York City in April 1925, leaders from the twelve major VNAs put the future to a vote.[23] The majority agreed that in the coming decade, care of the sick at home would be conducted by private agencies subsidized by public funds and that health departments would limit their work, more or less, to the exercise of police powers when necessary for the preservation of the community's health. The vote demonstrated visiting nursing's disdain for health departments and their ability to deliver nursing services. The VNAs believed they were the only legitimate recipients of public funds for all nursing care in the community.[24] In their opinion, they were the best providers of cost-effective community-based care and therefore entitled to municipal subsidies similar to those given to hospitals to care for the sick poor. The "prejudice" that led municipalities to underwrite hospital care while regarding home care as "superfluous" was beyond their understanding.

Of course, what the advocates of visiting nursing failed to grasp was the growing disparity in community support for home care versus hospital care. By the late 1920s, medical, surgical, and even obstetrical patients of all classes were seeking hospital care. With the circumstances that created a need for home-based care no longer of major concern to most communities, visiting nursing's original vision and mission seemed out of step with the times. As Lillian Wald prophesied in 1915, "It takes imagination to visualize the steady, competent, continuous routine [of the visiting nurse] so quietly performed, unseen by the public, and its financial support is even more precarious because there can be no public reminder of its existence by impressive buildings and monuments of marble."[25]

The growing volume of hospital care was impressive. In 1922, patients in the United States spent more than fifty-three million days in general hospitals—the equivalent of one day for every two members of the population. Hospital construction boomed in the 1920s; by the end of the decade, the capital investment in hospitals reached an estimated $3 billion, one of the largest enterprises in the country. Numbers of hospital beds were growing faster than the population. By 1930, the United States had 6,665 hospitals with more than nine hundred thousand beds. With total bed capacity increased 115 percent since 1909, the need for home care vastly diminished. The more successful the hospitals, the more they appealed to paying patients. Any value attributed to home care was based on its ability to relieve hospital deficits through care of nonpaying patients.[26]

For those who could pay, medical experts warned that conditions were no longer favorable to home care and recourse to the hospital had become necessary. The hospital was part of modern society. Tiny flats and small apartments did not accommodate the sick, and urban life was not congruent with long-term family caregiving. More women worked or were not at home to provide care. Hospitals were clearly the best place for surgical care and the technologies associated with scientific advances in medicine. As early as 1913, Henry Hurd, superintendent of the Johns Hopkins Hospital, had unhesitatingly predicted that "the day of general home care of the sick can never return. Social conditions forbid the possibility that hospitals will become superfluous."[27] Families and patients clearly agreed and increasingly chose hospital care. The dominant paradigm favored medical innovation and acute-care intervention. Insurance coverage and philanthropic enthusiasm were directed toward institutional care for expert diagnosis, treatment, and cure.

In the absence of widespread support from the community, the VNAs admitted that preference for hospital care affected contributions and financial well-being. Reluctantly acknowledging that in the future there would be fewer patients of any kind at home, some VNAs reconsidered their options. The need for financial sufficiency was paramount; the only option and the only obvious, immediate source of income was the paying patient. While the sickest of patients were now hospitalized, the VNAs understood that patients still required care in their homes—it was simply a different kind of care.[28]

"But You Thought a Henry Street Nurse Gave Free Service Only?"

That few people sick at home purchased the services of a VNA nurse suggested the basis for a mutually beneficial solution for patients and visiting nurses. What the middle class needed was intermittent, affordable nursing care. For the most part, all that was available was care provided by full-time private-duty nurses—beyond the means of most families. The solution was nursing care sold by the hour, but the VNAs quickly discovered that their hourly service placed them in competition with private-duty nurses also searching for a way to make a living. If hourly nursing was the salvation for both the visiting and the private-duty nurse, the next issue was how it would be organized, by the VNAs or by private-duty nurses' registries.[29]

For many boards of managers, the paying patient was as yet a largely uncultivated means of survival. According to Mary Gardner, three groups of patients should be reexamined by every VNA. The first class included the aged, the chronically ill, widows with children, and others who, because of health or other handicaps, could never be self-supporting. Most needy patients unable to pay physician or nursing costs were in this group. The second group consisted of those with reasonable means who could make a small payment for nursing services. Most associations decided very early on to make their services available to this class of "thrifty wage earners" who were unwilling to accept charity and from whom the nurses would willingly collect money. Although many agencies claimed this as the class of patients for which they obtained their best results, it was nevertheless difficult to collect payment. The third class of patients was those with incomes well beyond the poverty level, but still not enough to afford a full-time private nurse. This class, as yet virtually untapped, could certainly benefit from a nurse during illness and might even pay the full price of a visit.[30]

While realizing that an hourly service presented the most promising avenue for survival and expansion, the VNAs failed to pursue this potentially lucrative field with any genuine enthusiasm. To succeed, the service needed to be made available by appointment, with the same nurse seeing the patient for successive visits. A great deal of publicity would be required to ensure the community's use of this service, and associations needed to transcend their fear of losing their philanthropic status in the eyes of supporters.[31]

Much more important than any of the obvious constraints of hourly work was its different "character," which tended to be too much like private-duty nursing. From the perspective of the VNA staff, hourly patients were simply too well-to-do and their problems too chronic. The more affluent required a leisurely approach, were "more talkative and less teachable," and rarely thought of helping themselves. Some even asked the nurses to perform maid services. For the most part, the staff found hourly work a burden and the services proved demeaning for a well-trained nurse. Although the VNAs asserted that they were offering a service superior to that of the average private-duty nurse, they demonstrated little willingness to give any preference to the needs of their hourly patients over their other, more "needy" patients. Predictably, with VNAs initiating hourly services in such a manner as never to "encroach" on their regular services, the outcome was less than spectacular.[32]

Although the VNAs claimed to meet the needs of paying patients, their efforts were correctly criticized as superficial. In 1939, Michael Davis, social reformer and director of the Rosenwald Fund, declared in total exasperation that VNAs were twenty-five years behind the hospitals in their efforts to expand services to people who could pay their way. Hospitals were proud of the services they rendered to paying patients, but VNAs with similar social obligations were, from his perspective, timid about it. Hourly nursing was considered a secondary activity, administered more as a concession than as an opportunity. As a result, Davis concluded, hourly services tended to be too inflexible to meet the needs and demands of the people who should use them.[33]

Surveys in 1927 and 1931 confirmed that hourly nursing was indeed a lost opportunity for most VNAs. In the few organizations offering hourly nursing, it accounted for only 1 to 3 percent of the work. Requests tended to come from patients, not physicians, and the majority of patients requesting care were chronically ill. Lacking clearly defined organizational plans, funds for publicity, and staff support, these services were not a success. In contrast to Davis's analysis, the VNAs concluded that the failure of hourly nursing to meet the needs of patients of moderate means was probably the result of overestimating that need or, perhaps, not offering service in a form that more precisely met patients' requirements or wishes. No matter the cause, the end result was that the paying patient would never become more than a very limited source of funds for most VNAs. In the absence of other remedies, survival appeared dependent on selling the mission of the VNAs to private and municipal sources of financial support. With the onset of the Depression, the VNAs found themselves increasingly dependent on fees from insurance companies. By 1932, the associations in some larger cities obtained more than 60 percent of their income from insurance contracts.[34]

Their Changing World

Throughout the 1920s and 1930s, visiting nurses retained as their key fundraising strategy the traditional rhetoric of "their health is your health."

Their chosen sound bite portrayed illness as a menace to the community, the VNA as an investment in self-protection, and the visiting nurse as defender of the public's health. As late as 1934, the Henry Street Settlement's annual report reminded supporters of its role as "protector":

> In a very real sense these nurses stand between you and the physical and social ills that are ever likely to beset a large city, just as the police stand between the community and crime. Communicable diseases, untended and unchecked, move rapidly, and desperation enters the homes of the hard pressed as a companion of illness. Not only is an untended child with scarlet fever in a lower East Side tenement an implied threat to the cradle on Park Avenue, but his father, desperate from poverty and worry, embittered at the world and receptive to anti-social ideas, is an implied threat to the social order. By helping to care for the child and to prevent similar cases, the nurse also helps to check the spread of social ills. And there is nobody but the Henry Street nurse to do the job that she is doing.[35]

Using "menace to the community" as her primary raison d'être, Boston's Mary Beard ranked the nurse's work, in order of importance, as caring for people with acute and communicable diseases, maternity care, and care for chronically ill patients. Old images and historic missions no longer described the dramatically transformed world of home-based care. In reality, the causes of death and illness, as well as the work of the visiting nurse, had little in common with the origins of visiting nursing at the end of the nineteenth century. "Dangerously ill" patients were infrequent, and only a few patients at home had acute, much less communicable, illnesses. Maternity patients were steadily increasing, while the percentage of chronically ill patients remained stable. Ironically, absence of the patient at the time of scheduled visits had become a major problem. Beyond the symbolic connotation of the patient not being sick enough to stay home, at twenty thousand wasted visits per year these missing patients were an expensive problem.[36]

Such changes were not unique to Boston. Data from the *Philadelphia Hospital and Health Survey* and the VNSP annual reports present a similar picture. By 1928, the most obvious changes were in both the rates and the causes of death for Philadelphians. Death rates from all causes declined from 21.7 per thousand cases in 1888 to 13.3 per thousand by 1928. Death resulting from typhoid, diphtheria, tuberculosis, scarlet fever, whooping cough, and childhood diarrhea was less common; death rates from pneumonia, measles, and injuries remained fairly constant; and different types of illnesses were in the ascendancy—diabetes, heart disease, and cancer. The dreaded contagions of the past were being replaced by chronic illness.[37]

In 1928, 113 VNSP nurses cared for 37,048 patients, making 269,542 visits at a cost of $267,368.74 (slightly less than $1 per visit). In the *Philadelphia Hospital and Health Survey,* Haven Emerson wrote confidently that "many hospitals cannot match [this] in volume, results, or cost." Nursing services in Philadelphia had grown apace in "usefulness, quantity, quality, and public esteem."[38]

Careful analysis reveals that in Philadelphia and elsewhere, care for patients with chronic illness now consumed almost as much nursing time and cost 2.5 times as much as care for the acutely ill. The VNAs were finding at least 20 percent of their income committed to the chronically ill. Furthermore, quick disposal of these cases was a most difficult matter, while retaining them created cost/benefit dilemmas. Chronically ill patients consumed a great deal of nursing time with little hope of recovery, and with few alternatives for care. For most VNAs, the chronically ill were an increasingly burdensome and endless obligation. As the ladies of Charleston had documented more than a hundred years before, caring for the chronically ill was a vexing problem, best avoided or referred elsewhere, if possible.[39]

The rise in chronic illness and decline in acute care at home were dramatic but predictable in the light of changing morbidity patterns and the growing importance of the hospital as the locus for acute and surgical care. In their quest for the latest advances in medical technology and science, families increasingly sought hospital care for acute illnesses, while chronic illness became the focus of care in the home. Chronically ill patients provided slim promise of cure and were of "little scientific interest," making them particularly unwelcome in most hospitals. Special institutions for the chronically ill were too expensive for most families and usually had waiting lists of two to four years. Occasionally, the poor patient with chronic illness found custodial care in a public institution or home for the aged and chronically sick, but most were left to struggle as best they could at home. If they were lucky, they found a visiting nurse to help out for as long as possible.[40]

"When Is a Chronic a Chronic?"

By the end of the 1920s, visiting nurses defined the "chronic sick" as individuals disabled by disease for a period of three months and incapacitated for an uncertain period of time.[41] Specifically, these patients could not perform the activities of daily living and needed medical, nursing, and personal care. Most were cared for by family, and the absence of sufficient family caregivers made chronic illnesses problematic. In some families the sole caregiver was the breadwinner, too exhausted or unwilling to give sufficient help. Most challenging were households in which the patient was alone. No matter what the circumstances, the nurse rarely could simply teach the family of a chronically ill patient what to do, then quickly leave. VNAs thus pondered the fundamental questions of how many "chronics" needed care, how much care was required, and what else was available in the community.

Experienced in community-based care, visiting nurses easily hypothesized solutions for the intractable problems caused by chronic illness. These included (1) assistance with personal care by an aide or attendant under the supervision of nurses; (2) boarding out in private homes; (3) temporary hospitalization for special treatments or transport of treatments from the hospital to the patient's home; and (4) assistance to family breadwinners by subsidizing their earnings or providing a part-time attendant to help with care. Unfortunately, these straightforward designs for management of chronically ill patients were not compatible with available systems for delivering or financing care.[42]

As usual, the possibility of creating a practice equal to their ideals was a question of nurses' power within the health care system. Visiting nurses firmly believed that chronic illness was a community problem, but found no one willing to accept the responsibility. Providing care to chronically ill patients in ways that enhanced their dignity and self-respect generated little enthusiasm. Most critical was the matter of who would pay for what many considered palliative or unconstructive care.

In its desperate search for ways to pay for care of patients with chronic illness, the Boston CHA even convinced Sophie Nelson of the John Hancock Insurance Company to test nurses' ability to cure or retard the progress of chronic illness. Because the company knew little about the natural history and prognosis of chronic illness, a yearlong study was undertaken. From an insurance perspective, the question to be answered was whether a payment system could be established that was stringent enough to avoid paying for long-term personal care, elastic enough to care for patients with the potential to recover, and humane enough to cover the care of patients requiring skilled care to minimize suffering. Conducted in 1928, the study compared the outcomes of unlimited care provided by the CHA with the outcomes of care restricted by insurance coverage and provided by six VNAs. The results showed that limited and unlimited care of the chronically ill produced the same outcomes, suggesting that care of the chronically ill was a humanitarian rather than a scientific process. Thus, from an insurance perspective, the most "expedient" policy was to pay for very limited care for chronically ill patients. In matters of the bottom line, diagnosis and prognosis still dictated the amount of service covered by insurance companies. Nelson concluded that insurance companies would not carry the load of chronic illness alone. The chronically ill patient was a family and community problem.[43]

The Unseen Plague of Chronic Illness

Naomi Deutsch, director of San Francisco's VNA, reminded colleagues in 1928 that most community plans for care of the chronically ill delegated the responsibility to visiting nurse services. Care of the chronically ill patient was the new challenge of the VNAs and, as in the past, development of this service needed intensive study and required its own expertise. Over the next twenty years, home care and the fate of the chronically ill would become intertwined.[44]

Throughout the 1930s, Americans struggled with the economic hardships imposed by the Great Depression. Widespread unemployment, growing welfare rolls, rising demands for free care, and declines in charitable giving left voluntary hospitals and VNAs in precarious positions. For most VNAs, the Depression resulted in salary cutbacks, personnel reductions, and curtailment of basic services. Eventually, the inclusion of home care in the Federal Emergency Relief Act brought some assistance. Additionally, the New Deal's broadly focused social welfare agenda put into place a selective safety net and provided jobs, unemployment compensation, and public works programs.[45]

With its typical candor, and with funding from the Commonwealth Fund, the NOPHN paused in the midst of the Depression for a moment of self-scrutiny. The findings were, as one commentary suggested, not wholly flattering. The problems associated with community-based care were complex, and the agencies providing that care operated almost entirely in isolation from the overall systems of care. In general, any existing relationships with other social and health organizations were casual and haphazard, permitting gaps and duplications in services. Great variations across the country reflected different communities' peculiar adaptations to local situations and needs.[46]

A decade later, in 1943, a similar study funded by the Metropolitan Life Insurance Company found little had changed. Most communities were simply unwilling to finance care at home. Apathy, uncooperative attitudes, and uncoordinated health services were the norm. Believing it was not their responsibility, city and county health departments were unwilling to provide care for the sick at home. Organizations that in the past had provided home care were discontinuing these services. With mounting evidence that the emergency caused by World War II would produce shortages of physicians and nurses, overtax hospital facilities, and cause serious gaps in health services, health analysts assumed that many community health services would be inadequate to the increased demands.[47]

In 1931, nationwide, 1,570 organizations and 6,021 nurses provided care for the sick at home. By 1940, the number of nurses had begun to decline; with the closing of the Red Cross, tuberculosis associations, and MLI nursing services, the number dramatically decreased. In 1955, only four thousand nurses were engaged in home care.[48]

During the 1930s and 1940s, nutritional counseling, care of polio victims, programs of mental and social hygiene, and services to mothers and infants would come and go, but the complex and costly needs of escalating numbers of chronically ill patients remained a permanent concern of VNAs and an endless obligation for family caregivers. A joint forum of the American Medical Association, American Hospital Association, American Public Health Association, and American Public Welfare Association acknowledged the growing problems created by the chronically ill. The forum, known as the Commission on Chronic Illness, saw the solution as obvious— send a nurse.[49]

By the early 1950s, there was general agreement on the national need for comprehensive community-based services. Home care was the answer: it produced better outcomes and happier patients and improved medical economics. Nevertheless, difficult issues remained—about scope, need versus demand, and appropriate conditions for providing care at home. The consensus was for a totally new vision of financing and delivering health care. The really big question was, "Can it be done?"[50]

As Ernest Boas, prominent advocate for the chronically ill, argued, justice and decency demanded that communities provide for ailing members unable to help themselves.

> The existing facilities for the care of the chronic sick present a very confused picture—patients at home who should be in hospitals, patients in convalescent homes occupying beds needed for another purpose; a mad confusion of patients and institutions admitting them grudgingly, and having admitted them not providing the care that they need. It is a scene of great disorder. Public and private hospitals, homes for the aged, convalescent homes, nursing and visiting doctor services, aftercare agencies, agencies for sheltered work, medical social service departments, family service agencies—every one of them accepts with reluctance the burden of the chronic sick, and tries to shift responsibility to another agency which is no better prepared for the task.[51]

Admitting that no single approach could solve these problems, Boas thought every community should establish a policy for the care of the chronically sick, including a plan for the use of hospitals, custodial homes, outpatient clinics, and home care. Haphazard development needed to be replaced by consistent policy and responsibility centered in one place, so that those in charge had the authority and means to carry out a comprehensive program of care.[52]

Boas's vision disturbed the customary practice of medicine and was an affront to the notion of individual freedom and choice in medical care. Only a reconceptualization of the delivery of health services and identification of new sources of payment would bring an alternative system of care for the chronically ill. In the United States, this would remain an impossible dream, as families struggled as best they could, home alone with what Boas called the "unseen plague" of chronic illness.

The Trend toward Hospital Care

While providers, families, and policymakers called for a whole new system of care, the hospital maintained its uncontested role as the centerpiece of health care. World War II had contributed dramatically to the institutionalization of medicine, and by 1945 there were thirteen hospital beds for every thousand Americans. The Hill-Burton program, launched in 1946, added another $15.5 billion in hospital construction and produced 143,000 more beds. With improved facilities and better geographic distribution, a greater proportion of the U.S. population had gained access to hospital care. In 1958, one in every eight Americans was admitted to the hospital—twenty-four million in all. The reasons for this movement of health care to the hospital were numerous and complex. Physician and patient preferences, access to the benefits of science and technology, supply creating demand, the growing availability of hospital insurance, new pharmaceuticals and surgical techniques, and increased longevity of the population—all these factors contributed to this trend. By the 1950s, providers of home care were anticipating a very different future.[53]

CHAPTER 9

Home Care Becomes
the Fashion—Again

*B*y the 1950s, the ascendancy of chronic illness and the growing stature of the hospital dominated the nation's health care agenda. The work and financing of home care were increasingly overshadowed and recast by these realities. And yet, visiting nurses were "still knocking."[1]

Visiting nurse associations (VNAs) still found patients through a variety of referral sources, including families and physicians. By the late 1950s, however, half of all referrals came from hospitals increasingly eager to discharge chronically ill and poor patients. While the number of patients cared for at home declined, with the burgeoning of chronic illness, the number of home visits per patient increased. Simultaneously, the average number of visits per day, the hallmark of productivity, reached an all-time low and the cost per visit an all-time high. As deficits spiraled, home care struggled to survive. Chronic financial troubles and the impending loss of insurance payments (see Chapter 7) made retrenchment, cutbacks, and curtailment of programs common. Concurrently, job satisfaction and staff turnover among visiting nurses became major problems.[2]

Now Where Does the Money Come From?

"How the money came" also changed. Finding themselves at a crossroad, many VNAs persisted by patching together a variety of funds from local community chests, investment income, fees from patients, contributions from friends, union-sponsored health services, and public dollars from departments of welfare. Various voluntary organizations such as the Cancer

Society, Infantile Paralysis Association, and Arthritis Foundation sometimes paid for home visits. Occasionally, arrangements with hospitals for visits to recently discharged patients provided temporary sources of income.[3]

A comprehensive study of visiting nursing in 1954 disclosed that, in the aggregate, expenditures commonly exceeded income. Typically, the sources of income for VNAs were community chests (44%), patients' fees (16%), municipal dollars (15%), and private contributions (10%). Public funds were a growing source of income—mainly dollars from departments of public welfare and the Veterans Administration, which inaugurated its community nursing program in 1950—and in 1954 such funds helped support 86 percent of VNAs.[4]

"Is Prepaid Nursing Care Possible?"

From 1935 onward, organized nursing campaigned for the inclusion of nursing services as an essential component of any comprehensive medical insurance plan. Voluntary health insurance was still uncommon, but by the late 1950s the purchase of such insurance protection increased dramatically.[5] Despite nursing's early and persistent enthusiasm for health insurance, few insurance companies sold policies that included visiting nurse services. In 1947, 11 percent of persons eligible for medical care through their health insurance were also entitled to home care. Most purchasers of insurance were unaware of their visiting nurse benefits, however, and rarely exercised this option. Because few traditional clients of VNAs could afford to purchase insurance, its growing availability had little financial impact on the these associations.[6] In the absence of greater public demand for home care among nonindigent and insured populations, coverage for such care remained inconsequential. Nevertheless, VNAs persistently anticipated the arrival of universal insurance for home care. As the long relationship with the Metropolitan Life Insurance Company (MLI) came to an end on 1 January 1953, both sides hoped health insurance would fill the gap with new sources of support.[7]

Ironically, one of the most supportive articles ever to appear in the medical literature was written by two MLI physicians. In "Visiting Nurse Service: Community Asset for Every Physician," published in the *Journal of the American Medical Association* in 1952, the authors declared visiting nursing essential, an underdeveloped resource in prepaid medical plans, a defense against socialized medicine, and deserving a role commensurate with that of the hospital. The pending discontinuation of the MLI's visiting nurse service was not mentioned. Rather, the authors wrote of "the growing importance of home care of the sick." They reviewed the need for visiting nurse services, the organization of these services, and studies documenting cost effectiveness. Readers were assured that visiting nurses maintained close communication with and reported directly to the attending physician and faithfully carried out orders. The article concluded with glowing excerpts from the letters of patients and physicians. Visiting nursing could not have asked for better public relations—even if perhaps part of a propaganda effort to counterbalance the impact of the MLI's withdrawing visiting nursing from policyholders.[8]

As the MLI left the business of nursing, others were considering home nursing care as a potential health insurance benefit in lieu of hospitalization. From an insurance perspective, however, any such benefit needed to resolve several complex issues: cost, utilization patterns, uncontrollable demand for services, unfamiliarity with this type of service, and local unavailability of home care services. As always, the bottom line was how to determine the impact of a home care benefit on the cost of health insurance premiums.[9]

The Health Insurance Plan of Greater New York, a nonprofit corporation that provided health benefits to city employees with annual incomes under $5,000, was among the first to experiment with a home care benefit. This "wise" use of nursing care was considered successful, as nursing organizations were paid first by actual cost per visit and later through a capitated arrangement.[10] Across the country, other prepaid group practice or community-sponsored plans provided home nursing coverage when "ordered" by physicians.[11]

In 1957, Blue Cross reported the findings of a five-year study examining the impact of visiting nursing services following early discharge of patients from the hospital. In a near déjà vu of earlier assertions by the MLI, Blue Cross claimed dramatic savings were achieved when nurses went into the homes of persons recovering from acute illness. Patients were pleased, physicians expressed satisfaction, and Blue Cross requested that the Insurance Law be amended to authorize the provision of nursing services in and out of the hospital.[12] In Massachusetts, Blue Cross–Blue Shield also experimented with supplemental policies that provided insurance coverage for nursing benefits during prolonged illness. In this program, customary charges were paid to any VNA with a Blue Cross contract.[13]

Describing home nursing, organized home care, and homemaker services as new approaches to the provision of care, insurers continued to examine the extension of home care coverage to growing numbers of chronically ill and aged persons. Despite several highly publicized successes with visiting nursing, a study by the Health Insurance Association of America in 1959 found most home care agencies in a state of turmoil. Actuarial data were so few that it was impossible to determine the incidence, duration, or cost of home care. Once again, the vast variations in home care made any assessment of its potential to reduce health costs impossible. Only in cases of serious illness was home care found to be a reasonable use of insurance premium dollars. Insurance companies were advised to remain abreast of developments in home care, gain experience by experimentation, and gradually accumulate evidence.[14]

As experience with home care grew, the financial realities of home-based care also became clearer. As was increasingly obvious, patients' functional (physical and mental) ability, not diagnosis, determined suitability for home care. And given the variability of home care, any comparison of the daily cost of hospitalization with that of home care was fallacious. By the late 1950s, the cost implications of what is now called "case mix" were readily apparent. Like any form of care, home care confronted high costs when patients with complex and expensive problems formed a substantial portion of the agency's caseload.[15]

Strategies for Survival

Convinced that home care was an essential component of any comprehensive health service, visiting nurses were dismayed by the failure of the insurance industry to recognize and underwrite such care. Equally disappointing was the public's indifference and failure to give even a "passing thought" to the value of including home care services in health insurance plans. The VNAs maintained a facade of optimism, but failed to produce a solution for growing deficits.[16] Replacement of injectable medications with oral preparations, the continuing large numbers of expensive "patient not home" visits, the decreased demand for bedside care, and the poverty that prevented most VNA patients from purchasing insurance coverage—all were among the explanations for VNAs' ever-present financial woes.

The VNAs examined the cost of visits by type to discover "how much for what." No agreed-upon methods for determining cost were ever established, however, so comparisons between agencies were as impossible as in earlier decades. Declining productivity was a worry, so staff examined how they spent their time. Their conclusion was that visits to chronically ill patients simply required a great deal of time, justifying the decline in productivity as within reasonable limits. The high cost of staff turnover and orientation of new employees and the effort required to "cope with the paper snow storm" that accompanied numerous special projects, contracts, and third-party payments were also evaluated and deemed inevitable.[17] Endless self-analysis notwithstanding, VNAs remained convinced that the need for nursing care in the home had not diminished and they found "no reason to change the basic objectives." Rather, they would continue to adapt programs and develop new relationships in order to achieve their objectives more effectively.[18]

Deeply committed to the community and confident of the need for their services, VNAs searched for a new strategy to sustain their traditional mission in these very different times. They revisited the idea of less expensive workers, now called "home health aides," even making them available twenty-four hours a day. Convinced that the nursing and medical staffs of hospitals needed more information about home care, VNAs sent "integrators," or liaisons, to advise, consult, and select patients for referral. Wishing an end to what one visiting nurse called "the patient goes home from the hospital—so what!" attitude, visiting nurses streamlined referral forms and regularly met with hospital nurses. Their goal was to "better bridge the gap between hospital doors and the fireplaces of patients."

The most spectacular of the various survival strategies attempted by VNAs in this period were their efforts to expand and intensify the services they offered. These new programs, begun by nine VNAs between 1949 and 1963, provided coordinated comprehensive home care. The first, Philadelphia's Intensive Home Care Plan, opened in 1949 as an experiment in the care of the chronically ill. Philadelphia's program was designed to offer patients at home the services usually available only in hospitals. The hope was that hospitalized patients could go home sooner, recovery would be hastened, and needed hospital beds would be released earlier. After three years of operations, the program maintained a caseload of sixty-three to one hundred mostly elderly patients. Although Philadelphia's Visiting Nurse Society (VNSP) claimed the experiment a success, it also discovered that coordinating multidisciplinary care was time consuming and twice as expensive as its traditional visiting nurse service. It ended its first year with a deficit and by 1951 was desperate for financial support.[19]

Another popular innovation was the amalgamation of all community-based nursing programs into a single "combination" organization. Typically, this involved a partnership between VNAs and city health departments, a merger of voluntary and tax-supported agendas. VNAs campaigned for this organizational restructuring, first recommended by the Goldmark study in the 1920s. By 1931, eighty-nine combined organizations were in existence, but only six survived the Depression.[20] The idea gained favor again during World War II as a strategy for a more effective use of scarce resources. By 1962, fifty-four combination agencies employed 1,506 nurses; by 1968, one hundred employed 2,269 nurses.[21]

These incredibly complex organizational structures made possible, at least in theory, a more complete spectrum of services, while conserving administrative costs and using personnel more efficiently. Philadelphia again led the way, hoping to be "of substantial help to the people of a troubled modern city." Following years of "dreams, strife, and negotiations" and helped along by a fortuitous political upheaval and reform in city government, Philadelphia combined its municipal and voluntary nursing services on 1 May 1959. This visionary act was fraught with difficulties; not until 1964 did it bring together the caseloads of an entire city.[22] Despite these innovations, the VNSP barely survived the 1950s and 1960s. While it claimed service as the reason for its existence and the program's scope and quality as its measure of success, money remained a vital concern.

Philadelphians were not alone in these struggles. Whether home care (when available) emanated from new types of agencies or from old-fashioned organizations, the battle to survive was always the same. Many hoped the growing availability of health insurance would relieve their financial worries, but the effect was only a "very small ripple." Nationally, less than 1 percent of income came from insurance payments.[23] In the meantime, the vast majority of home care remained the responsibility of family members. As in the past, their efforts were guided by the newest and most authoritative of the numerous guides to home nursing.[24]

Home Care Becomes the Fashion

According to U.S. Surgeon General Leonard Scheele, writing in 1955, "During the past ten years, the increasing number of persons requiring long-term care, the high cost of institutional care and of hospital construction, and a growing awareness of the adverse effects of prolonged institutionalization have stimulated a keen interest in the provision of care to patients at home."[25] Simultaneously, the U.S. Public Health Service and the Commission on Chronic Illness were inundated by requests for information on home care programs. As if decades of service by VNAs had never occurred, the government and the American Medical Association "studied" home care and prestigious medical journals heralded its coming of age. According to Harvard professor Franz Goldmann, home care was "becoming respectable, if not fashionable."[26]

With physicians at the helm, the newly reinvented home care movement was proclaimed a dynamic approach to the far-reaching problems of "the chronic." Chronic disease, still described as "a hidden and insidious plague," was now responsible for more than half of all deaths in the United States. The nature of chronic disease—obscure in origin, of long duration, and usually occurring at an advanced age—made it both a social and a medical problem, yet its variability suggested no simple solutions.[27]

From 1930 on, needed improvements in the care of the aged were framed as a "problem of individualization"—suddenly home care promised a modern solution for a chronic dilemma and a cost-effective alternative to hospital or nursing-home care. *Home care* was no longer a generic term denoting the place where care was given—now it required a clear definition and specification of services covered.[28] Its true contribution would be "development and activation of an individualized concept of medical care, socially oriented and administered by a cohesive and informed medical team."[29] The underlying philosophy of this new care at home encompassed medical, social, and economic considerations. It claimed a unique modus operandi for referrals, patient selection, visiting patterns, use of auxiliary personnel (read "visiting nurse"), medical records, and hospital-like services and technology.[30] In recognition of its growing presence as a topic in medical journals, "Home Care" appeared as a heading in *Index Medicus* in 1953.

Although nurses had discussed and written about home nursing service since the late nineteenth century, the profession saw merit in a revised language and more comprehensive approach. As one visiting nurse suggested at the 1952 meeting of the American Public Health Association, "Perhaps we have at last found in 'Home Care' a term which can be used in all programs . . . this is good reason to make the term popular."[31] Despite the enthusiastic rhetoric of the 1950s and the unexpected imprimatur of physicians, the definition of home care was hardly innovative. It was still a program for "selected patients [who], while homebound, were provided with a full range of services, arranged for and coordinated through one administrative agency or organization."[32]

The operative phrase was, as it always had been, *selected patients.* The origin of this hospital-based back-to-the-home movement is generally attributed to the Montefiore Home Care Program, which opened on 1 January 1947. Acknowledging that the concept of home care was not new, medicine was most enamored with the "completeness and success" of Montefiore's program. As an editorial in the *American Journal of Public Health* asserted in 1949, Montefiore was seen as a hospital truly "moving into the future." Predictably, the physician-inventor of the Montefiore program, E. M. Bluestone, claimed his program as the "parent home care program," the first permanent, organized example of home care. He characterized earlier experiments in home care as limited and lacking—a lukewarm approach to the problem. Because they were not integral to the hospital and lacked total medical control, they were "haphazard" organizations without the power, authority, or discipline essential for optimal home care.[33]

Coordinated Home Care Programs

Descendants of the Montefiore model were commonly referred to as *coordinated home care programs* (CHCPs). The services provided varied with the program's purpose, method of administration, and type of patients served. All included nursing and medical care. Most provided medications, medical supplies, and appliances. In addition, social services and physical, occupational, speech, dietary, and rehabilitation therapies were available and, in some cases, transportation, x-ray and laboratory services, meals, and homemaker and housekeeping services. Occasionally, chauffeurs were provided so that physicians could efficiently complete their forays into the community. By the 1950s, home care programs claimed that even paracentesis, infusions, and blood transfusions were offered at home.[34]

Like their turn-of-the-century predecessors, these programs were portrayed as experiments "based on variations in community needs and resources." Most patients were medically indigent, chronically ill, and female. "Patient stay" differed markedly from program to program, ranging from six to twelve months, and the number of patients served ranged from a few to thousands. Likewise, patient age had no discernible pattern. Programs were administered by hospitals, physicians, VNAs, health departments, city welfare departments, and private social agencies, in one case by an independent committee, and in another by a medical school. Finally, the cost per patient or per patient-day of care varied widely from one program to another.[35]

Among its many promises, coordinated home care claimed it would bridge the gap between in-hospital and outpatient care, meet the needs of the chronically ill who did not require institutional care, shorten length of hospital stay, reduce hospital readmissions, offer the homebound patient coordinated and comprehensive care, and offer the family physician a substantial array of services. It was a dynamic program of medical management, involving the home, the outpatient department, the hospital, and other facilities. The poor were provided with "a quality of care not normally available to them, and at less cost."[36]

Hospitals showed the greatest interest in establishing these "new" home care programs. Administrators and staff shared an almost "universal agreement" that employment of such programs reduced the costs of hospital care, with home care (once again) seen as a means to reduce the hospital deficits caused by caring for the poor. According to one hospital spokesman, home care would save millions in capital outlay, as occupied beds could be emptied sooner and waiting patients admitted, and no additional expenses need be incurred for expanding bed capacity.[37]

CHCPs allegedly cut per diem hospital costs by 25 percent. The only study that carefully examined this question of savings, however, found some reductions in use of hospital beds but less than optimal results in total home care patient-days. The American Medical Association found no direct relationship between cost and number of patients served. Few acknowledged a fundamental reality: any economic advantages of home care probably derived from the unpaid contributions of family members—primarily women. Dr Sidney Shindell of the U.S. Public Health Service was the one exception, recognizing that "care in the home involves the assumption on the part of the patient's family of all of the domiciliary functions and *cost*, as well as a large percentage of the nursing functions and *cost*. It is in these categories that much of the savings occur."[38]

Of course, only in a fragmented system of health care could the shifting of cost be considered a measurement of savings. Some argued that reduction in the average length of hospital stay by early discharge to a home care program did not equate with savings in hospital operating costs. Rather, the opposite occurred: the average cost per patient-day tended to increase as the average length of stay decreased. In the absence of any reliable data on cost, incidence, or duration of care, it remained unclear whether home care did or did not reduce health care costs. Despite only "highly theoretical" knowledge of home care costs, the promise of reduced hospital stays remained an intriguing cost-saving strategy, deemed worthy of consideration and experimentation. Organized home care for the chronically ill, disabled, or convalescent also counteracted what was considered an overemphasis on institutional care.[39]

Enthusiasm did not diminish the fact that "certain difficulties arose" when home care was put to the test. Most persistent was the antagonism toward home care among visiting house staff. Attending physicians feared that extension of hospital services into homes was a step in the direction of socialized medicine. House staff feared that additional work and time devoted to home care would not receive specialty board approval. As one Boston physician admitted, visiting ten patients in their small walk-up apartments taxed even the most committed physician's strength and time. Predictably, hospitals had difficulty in obtaining qualified young physicians to take positions in their home care programs.[40]

With the exception of the indigent, the savings from shortened hospital stays—or no hospitalization at all—accrued to the patient or community, not to the hospital or insurance company. An estimated 15 to 20 percent of hospital patients with acute illnesses qualified for home care. Although home care was assumed to be less expensive per diem than hospital care, it was also understood that maximum use of such care among nonindigent persons required the creation of a broad system of home care programs and appropriate insurance coverage. As early as 1957, those engaged in the home care debate were cautioning that programs might be less effective in reducing hospital bed utilization than had been suggested and that "such programs should be regarded primarily as a *qualitative addition* to community health services." While organized home care might save as much as 50 to 75 percent of the per diem cost of hospital care, it seemed to have about the same per diem cost as nursing-home care.[41]

Health insurance companies already provided coverage for some home care services. From a voluntary health insurance perspective, expansion of organized home care coverage to a larger nonindigent population presented several problems. Organized home care programs remained experimental and were not uniformly available. Furthermore, the lack of reliable, usable data on cost, incidence, and duration made home care a highly speculative investment. Too little was known about what caused wide variations and fluctuations in cost. The many variables resulting in differential utilization of services among patients clearly needed investigation, along with the impact of disease category, age and sex of the patient, family structure, and cost differentials between home and institutional care. Insurers worried that reimbursement methods might create their own perverse costly incentives. They wanted control of coverage so that misuse, abuse, and overuse would be kept to a minimum.[42]

Inclusion of social and other nonmedical care also presented problems for companies offering health insurance, including any social welfare services appended to medical programs for the sick at home. As one insurance expert asked, since certain forms of home care included services traditionally outside the realm of medical care (e.g., homemaking, social work, occupational therapy and vocational counseling, health education, and nutrition services), to what degree were health insurance companies obligated to cover the costs of such services? Clearly, the challenge was to confine health insurance to payment for medical care and to avoid payment for personal or custodial care. After much study, the Health Insurance Association of America concluded that there seemed little "relationship between organized home care programs and voluntary health insurance." Although this was a reserved rebuke by the insurance industry of coordinated home care programs, home care's supporters would later claim their experiment as the "prototype" for Medicare's vision of home health care.[43]

New Era of Home Care

By 1964, only seventy CHCPs were in effect, scattered across the nation. Each was an intensely local undertaking, providing a locally acceptable solution to a particular community need. National distribution was uneven. Very reminiscent of turn-of-the-century experiments in visiting nursing, most CHCPs (fifty-four) were located east of the Mississippi and most (sixty) were in major cities. Establishing a community resource, caring for the homebound, releasing hospital beds, and complementing family physicians, medical education, and research continued as goals. CHCPs were generally small undertakings; overall, they served only fifty-five hundred patients in 1964.[44]

Although many programs admitted paying patients, few such patients sought the services of a CHCP. As always, money flowed in from various combinations of private, public, and philanthropic sources that were remarkably eclectic and geographically variable. Payment sources included city, county, and state departments of health and welfare; grants from the U.S. Public Health Service; prepaid health plans and insurance contracts; various foundations, societies, and charities with health concerns; hospitals, medical centers, and medical schools; united/community funds; private donations; and patients' fees.[45]

In spite of all this professional activity, families remained integral to home care's ability to fulfill its promise of better care for less cost. With home care, it was argued, patients would have a more satisfactory recovery and families would have the "satisfaction" of caring for their own. Every American, "whether he is housed in a well staffed mansion or in a bleak furnished room," could, with proper community planning and organization, receive high-quality or adequate care. Some acknowledged that not all families were willing or able to accept the burden of providing home care. Often, conditions for home care were far from ideal, and families needed sensitive caregivers to help make the home environment suitable. Family members and friends also had to be taught how to provide care and often needed assistance with housekeeping.[46]

The prevailing view was that, despite these difficulties, by giving families who were willing and able (physically and psychologically) the right combination of care at home, a filled hospital bed could be emptied. Economy and efficiency came with the "new look" of physician-directed, centralized, coordinated, and comprehensive care for homebound patients with "complex" problems. Even the American Medical Association endorsed home care as "the most satisfactory means of meeting the total care needs of certain patients or a patient during certain phases of his illness." From the association's perspective, hospital involvement, broad community interest, and the support of the medical profession had finally allowed home care to take its place as an integral part of the continuum of patient-centered care.[47]

Rhetoric and hyperbole notwithstanding, *coordinated home care* became part of the new vocabulary of health care policymakers searching for potential cost savings. In the light of this newfound interest in home care, the small numbers of patients in CHCPs, lack of a dependable formula for calculating cost, and only the vaguest sense of how to select "suitable patients"—along with the vast variation in what we now call "resource consumption"—were mere details to be worked out over time. In the meantime, the majority of home care patients remained the responsibility of visiting nurses across the country.[48]

Services Available for Nursing Care of the Sick at Home

The increasing interest in establishing and expanding home care programs also created a new demand for more current information about the availability of these services nationally. Beginning in 1959, the Division of Nursing of the U.S. Public Health Service began to publish data on the availability of nursing care for the sick at home. In 1963, 1,163 agencies provided some form of home care in 2,323 communities, making care available, at least in theory, to 55 percent of the U.S. population. At the same time, less than 1 percent of the population received continuing nursing care at home. Not surprisingly, most (692) of the agencies delivering home care were VNAs, and of the remaining programs, nearly one-third (384) were health departments, 6 percent (85) were combined government and voluntary organizations, and only two were hospital-based programs.[49]

The available data suggested a new interest on the part of health departments, no doubt economically motivated, in caring for the sick at home. In the early 1960s, three developments enabled health departments to acquire sufficient funds to provide home care. First, with passage of amendments to the Social Security Act, health departments could receive federal funds earmarked for Old Age and Medical Assistance. Second, in 1961, the Community Health Services and Facilities Act allowed health departments to collect fees for services provided under this legislation. And third, health departments could contract with insurance companies, welfare agencies, and other organizations and collect fees for "comparable service." By 1963, 261 official health agencies were in a position to provide home care and collect fees from Old Age Assistance, other public assistance, Medical Assistance for the Aged, the Veterans Administration, Blue Cross, voluntary agencies, and industry. Having transcended the legal obstacles to accepting fees, health departments were suddenly no longer reluctant to provide "curative services." Such a philosophical shift, however, meant that public health nurses would be returned to the bedside, and this required more than "conviction and understanding" on the part of presiding health officers. If health departments were to provide satisfactory care to the sick, nursing staff needed refresher courses, intensive in-service education, and supervision by skilled clinicians.[50]

"*I Am More at Home—at Home*"

Reporting on New York City's "celebrated experiment in home care," statistician Mabel Reid described the typical home care patient as a sixty-four-year-old woman with heart disease, living with her family without public assistance, and unknown to the visiting nurse prior to referral from the hospital. This patient received forty-two nursing visits a year at a rate of 5.7 visits per month. Approximately 2 percent of the time, the patient was not at home when the nurse visited. Visits were made weekly, but occasionally daily visits were required. The average visit lasted thirty minutes.[51]

Home visits made by most visiting nurses involved providing care or giving a treatment, anything from assistance with transfusions to injections, irrigations, urinalyses, enemas, changing dressings, preventing bedsores, removing sutures, applying ointments or eye drops, shampooing hair, and checking blood pressure. Teaching or supervising family caregivers and finding ways to lighten the burden of caregiving were of prime concern and required ingenious solutions. Patients' attitudes were also a major concern. Patients requiring help were described as ranging from helpless to uncooperative. Some exhibited fear and extreme apprehension, depression, or hopelessness. Patients' poor understanding, motivation, and willpower were the home care nurse's constant challenge.[52]

Limitations in physical ability and the need for assistance with the routines of daily living were common problems reported by most home care patients. In established home care programs, the team approach was not just a "pious avowal." Depending on patients' needs, they received visits from nurses, physicians, social workers, and various therapists, and received housekeeping services. The typical home care patient in 1958 was cared for by two or three members of the team; with few exceptions, one of these team members was the physician. Despite such a rich array of services, patients frequently were readmitted to the hospital.[53]

Inventing a Federal Policy on Home Care

Both the history of national health insurance and the role for federal financing of health care have been extensively chronicled. In a substantive analysis of public policy and home care, A. E. Benjamin convincingly argues that these policy developments have been shaped by (1) the characterization of home care as a residual set of services for patients whose conditions are not amenable to mainstream medical intervention; (2) an absence of consensus about home care's principal goals, essential elements, and place in the continuum of care; (3) the presence of two paradigms of home care—the easily articulated medical-postacute model and a more nebulous supportive social-welfare model; and (4) the hinging of home care's legitimacy on its ability to relieve the utilization and costs of institutional care.[54]

Given home care's variability, marginality, and ambiguous costs and benefits, its inclusion in any federal health program would seem unlikely. Yet home care was first included in federal legislation (the Wagner-Murray-Dingell bill) in the series of unsuccessful proposals for compulsory health insurance during the Truman administration. Defeat of this legislation did not defeat efforts to achieve at least an incremental form of national health insurance. Throughout the 1950s, various proposals dealt with financing medical care, with an increased focus on the elderly. In 1957, Representative Aime Forand organized the next serious push, but this time home care was missing.[55]

Although the Forand bill never moved out of the House Ways and Means Committee, it attracted wide media attention and rekindled public discussion of the aged and their health problems.[56] In 1959, the Joint Economic Committee of Congress reported that "a revolution of rising expectations is in progress in the field of health and medical care." These concerns became strikingly evident as Senator Pat McNamara's Subcommittee on Problems of the Aged and Aging held hearings around the country. A member of McNamara's staff later recalled that they knew the issue was popular, but no one ever realized how popular. Throughout the summer and fall of 1959, in city after city, the experts would speak, and then the microphones were turned over to the audience. "The old folks lined up by the dozens" and talked mostly about medical care. "For the first time, we had these people telling what life was like for them—and letting us know they were more than a lot of statistics." The hearings produced headlines and front-page stories. Critics claimed the hearings were politically inspired and the resulting report failed to represent the views of the elderly. But transcripts of the hearings, as well as the joint committee's report to the Senate, did in fact urge change. The final report was released in February 1960, with the first recommendation to expand Social Security "to include health service benefits for all persons eligible" under it.[57] As a result of the 1959 hearings and the subcommittee's survey of problems and potential solutions, home care reappeared as an essential health service.[58]

Testimony on behalf of home care was extensive, dramatic, and irresistible. Home care was important and humanitarian, a simple, commonsense, community-based service that preserved home life and was "enormously economical."[59] Recommendation I of the Senate Committee report, *The Aged and Aging in the United States: A National Problem,* concluded, "The prevention of illness, its early diagnosis, and restoration to health are primary goals of an adequate health service. Diagnostic services and home care programs under supervision can be effective in reducing the high cost of hospitalization and institutionalization. A new emphatic approach to organized home care services, particularly for chronic illness and for preventive efforts, may reverse the rapidly rising cost of hospitalization, as well as improve the health of America's older citizens."[60]

In the judgment of the Joint Economic Committee, home care had already achieved impressive results. Despite the shortcomings of existing programs, home care's potential for solving many problems "inherent in haphazard institutionalization" of patients with long-term illness warranted its extension and continued study. Organized home care was a means of providing a variety of appropriate services in a coordinated fashion to patients who were sick at home. The nucleus of the home care team included a physician, nurse, and medical social worker. Other services (physical therapy, speech therapy, homemaking, vocational counseling, and loan of equipment) could be added as needed. Finally, the committee acknowledged that "the cooperation of the patient's family and the suitability of the home environment are important factors related to the success of such programs." While recognizing that home care programs were costly, the committee concluded that from a "dollars and cents" point of view, home care's restorative and preventive emphasis promised to cost less than programs that increased numbers of hospital or nursing-home beds.[61]

The eventual outcome of these various legislative efforts was passage of the Medicare legislation in 1965; the complete political details are well told elsewhere. How home care came to be part of Medicare, however, bears retelling.[62]

Backed by popular demand and endorsed by the Subcommittee on Problems of the Aged and Aging, inclusion of home care in the legislation was almost assumed. McNamara's hearings were followed by the White House Conference on Aging, which brought together hundreds of authorities in the field; the conference participants also endorsed home care as an essential component of health services for the aged.[63] Despite variations on the general theme, home care was no longer an afterthought and appeared in all subsequent bills leading up to eventual passage of the Medicare legislation.[64] Congressional testimony between 1961 and 1965 reviewed the familiar themes and debates associated with home care. The American Nurses Association and the larger VNAs supported any legislation that provided more home nursing care. The American Medical Association emphasized medical involvement and oversight. Social service interests stressed collaboration and the importance of supportive services. Ever-skeptical representatives of private insurance companies remained cautious and tentative, warning of the often unintended and expensive consequences of home care. In any event, from 1960 onward, home care appeared in every proposed bill.[65]

Some contend that decisions about the home care provision were made by a small cadre of Social Security Administration officials who led the fight for the King-Anderson (H.R. 3920) legislation, the official proposal of the Kennedy-Johnson administration. Claiming inspiration from the CHCPs, especially the Montefiore program, these officials also acknowledged substantial assistance from Blue Cross. Others suggest that a well-structured and easily understood cost-accounting method for home-based services provided by the National League for Nursing shaped the financial aspects of the home care legislation. The league staff member had had previous experience at the MLI, and her cost-accounting method was said to be so clear that the "Feds felt like home care agencies—unlike hospitals—were not trying to pull the wool over their eyes through some fancy accounting methods."[66]

As the debates over publicly financed health care continued, worries about uncontrollable costs mounted. In a now familiar chorus, some warned that the home care benefit contained great possibilities for expansion and substantial increases in cost, while others promised it would save money. In the end, the provision of home care was a political strategy, more symbolic than emblematic of true budgetary considerations. The perceived capacity of home care to empty hospital beds destined it to become a post-hospital benefit in the Medicare program. Following a hospitalization of at least three days, patients could receive up to one hundred home care visits. Home care would be supervised and planned by the patient's physician. Home care patients were expected to be homebound and to need only intermittent care. Nursing care, therapies, and part-time home health aides were among the services provided. Payment was based on "reasonable cost" incurred in providing care for beneficiaries. In acknowledgment of the importance of a restorative and comprehensive focus, the care provided at home by the Medicare program was now called *home health care.*[67]

"Bracing Ourselves for the Advent of Medicare"

On 4 November 1965, on the eve of the enactment of Medicare's home health benefit, Elizabeth Madeira, president of the VNSP, presented her annual report.

> Each year at our Annual Meeting it is our duty and privilege to pause and take stock of what we, as the Board of Managers of the Visiting Nurse Society of Philadelphia, have accomplished during the past year and to consider how we should plan for the future; to weigh our worries and frustrations of the moment against the goals achieved and the care and service which our staff has been able to give to the people of Philadelphia. At this moment in time the worries and frustrations loom large.[68]

The plight of the VNSP was caused by a $96,000 deficit, with undesignated principal being spent at an alarming rate. In addition, there was a "spate of resignations" among top-level nursing staff, just as the Medicare home health benefit was being enacted. After expressing amazement at the board's courage to choose her as its president, Madeira prognosticated that "in spite of our problems, I think we can face the coming year with confidence." Given the circumstances, one wonders if such fortitude was a peculiarly Philadelphian attitude or simply a prerequisite for the leaders of any organization attempting to provide care for the sick at home.[69]

Across the country, implementation of the Medicare program on 1 July 1966 marked the beginning of a new era for home care. Reviewing home care's past, present, and future, Claire Ryder, a Special Assistant for Continuum of Care in the U.S. Public Health Service and a prime mover in implementation of the Medicare home health care benefit, declared that home care was now on the move—it was a "happening," taking its rightful place in the continuum of health care.[70]

These new government-sponsored home care programs were financed primarily (51%) through the federal Medicare program, but three additional federal programs also provided community-based services for the poor of any age and the elderly: Medicaid, Title XX of the Social Security Act, and Title II of the Older Americans Act. Monies for most of these programs were to be collected by the federal government and redistributed to individuals through intermediaries to subsidize the purchase of home care services in the private market. However, while the Medicare legislation mandated the kind and duration of services, the other federal programs allowed more state and local discretion in the services provided. Consistent with American tradition, this federally financed and regulated pluralism prohibited the development of a coordinated system of home-based care. Instead, the outcome for many patients was a confusing assortment of programs with different payment requirements, eligibility criteria, and systems of reimbursement, as well as coverage limitations that created gaps in services.[71]

Ryder's "happening" looked remarkably familiar. Although it was now called home health care, it still assumed numerous forms, operated under a variety of administrative auspices, and displayed impressive regional differences. Nursing services were the "keystone" and physician involvement was rarely extensive. Most home care agencies were small, with 43 percent reporting a staff of only one or two nurses. Three out of five agencies were official health departments, one in four was a VNA, and one in thirteen was administered by a hospital. At the onset of the Medicare program, 1,356 agencies were certified to participate. Growth and expansion were essential if home care was to fulfill its promise. By 1969, the number of participating agencies reached 2,184. In 1967, Medicare's home care outlay was $46 million.[72]

Despite this infusion of federal funding, the complex problem of caring for the sick remained unresolved. As researchers and medical economists of the 1960s examined these problems, they asked the standard questions: Was home care cost effective? Which problems were best treated at home? What type of home environment was required? Were physicians and patients willing participants in home care? And what was home care's place within the matrix of medical care? Like their predecessors, they began the dialogue confident that in the future they would be able to clarify and answer these perplexing questions.[73]

Almost overnight, Medicare redefined home care to include only those selected functions and prescribed circumstances that were reimbursable. This time, a federally sponsored insurance system sought to establish home care as an alternative to institutional care. Once again the payment system created a narrowly defined, fragmented, and uncoordinated set of acute-care services ill adapted to the needs of the chronically ill at home. As Brahna Trager pointed out in her 1980 policy analysis, Medicare's major emphasis was on service selection and delivery patterns that most closely resembled institutional care.[74]

Despite this infusion of new funding, home care quickly demonstrated that it remained a less than perfect solution to the health care needs of an aging, chronically ill population. Home health services were an ineffective alternative to institutional care. Family caregivers did not materialize beyond the "wishful thinking" of policymakers. Development of home care agencies offering a full range of essential services was disappointingly slow, and in many communities nonexistent. Where available, home health services were underused. Providers found that the limitations, inflexibility, and complexity of the new laws and regulations made it impossible to provide services in the usable form required to ensure a patient's "life in the community." The maze of entitlements, functional eligibility requirements, and continuous shifts in reimbursement sources proved self-defeating for a policy created to encourage community rather than institutional care. Most ironically, Medicare home health services for people who were virtually housebound and had severe functional limitations did not include maintenance of their environment. Such services were considered inappropriate for a health insurance program because they too closely resembled the financing of a household servant or maid. The reluctance did not bear up under rational examination, for as one author suggested, "A clean body in a dirty house may not be good health care, but it is what we pay for."[75]

The failure of the Medicare home health care benefit to provide the essentials of daily living, nourishment, and a safe and hygienic environment made it an unreliable resource for many. Although Medicare expenditures for home care grew, many families of the elderly and their physicians found home care unfeasible and institutional care much less trouble. As always, the majority of home care came from informal support systems—family, friends, or neighbors. Policymakers and third-party payers who considered home care of little intrinsic value found the evidence of its cost effectiveness inconclusive and its benefits difficult to control.[76]

EPILOGUE

The Future of Home Care

*I*n November 1999, after a decade of research for this book, I faced a difficult task. During those years of research and writing, I had often stared at the message on an orange index card posted above my desk: "To the visiting nurses of Boston who for fifty years have cared for the city's sick with an unchanging purpose in a changing world." The message is from a plaque presented to the Boston Visiting Nurse Association in November 1936. I had titled this index card "The Problem?" suggesting to myself the hypothesis that failure to change had relegated an essential and sensible method of caring for the sick at home to the margins of our health care system. As I was bringing this project to a close, I found my hypothesis was wrong and my conclusion, at best, disheartening. Home care both held fast to its original purpose and responded to a changing world; care providers retained definitions and objectives while simultaneously making necessary compromises and adjusting methodologies. Despite this clarity of vision and resourcefulness, home care never lost its vulnerability. Nor has it yet found its place in modern health care.

The Legacy of Home Care

As chronicled in this book, organized home care in the United States began in the antebellum South. The ladies of Charleston, South Carolina, developed a system of caring that was compatible with the needs of the poor and the southern way of life at that time. Domestic expertise and antebellum notions of civic and religious duty came together to produce an enduring set of themes central to home care: mission and money. The ladies of Charleston were also the first to document home care's "vexing" predicaments of family circumstances, race relations, and chronic disease. As they quickly learned, families and their home lives were unpredictable and often uncontrollable, yet they were vital determinants of the outcomes of care. From the beginning, this was a women's story, since both patients and caregivers were almost exclusively women.

Sporadic efforts by social reformers and religious groups followed, but it was entry of the trained nurse into the homes of the poor in the 1880s that transformed caring for the sick at home. Home nursing care of the poor was financed by philanthropists and provided by a visiting nurse. The ailments encountered by these nurses were often acute, frequently complex, and almost always complicated by difficult social and economic circumstances. The visiting nurse's mission was to care for the sick, to teach the family how to care for its sick family member, and to protect the public from the spread of disease through forceful yet tactful lessons in physical and moral hygiene. Medical care of the urban poor varied with the circumstances of the patient, ranging from visits to dispensaries or folk healers (or both) for most ailments to house calls from private physicians in times of acute crisis. Hospital care was sought only in the absence of all other alternatives.

From its inception, home-based care differed for the rich and the poor. For the middle and upper classes, those who could afford to purchase care, illnesses were usually supervised at home by the family physician. Even the most seriously ill were often treated at home; the patient's bedroom was the workplace for most nurses and physicians. Nursing care was provided by a private-duty nurse who remained with the family for the duration of the illness and received fee-for-service payments. One of the greatest difficulties with this approach to privately purchased care was its unpredictable, individually driven system of distribution. Demand for these nurses was sporadic and seasonal. By the beginning of the twentieth century, the supply of private-duty nurses exceeded demand.

As knowledge of the work of visiting and private-duty nurses spread, the number of organizations providing such services rapidly expanded. In many U.S. cities, this expansion was characterized by an idiosyncratic mix of government, voluntary, and entrepreneurial initiatives. Despite tremendous growth overall, most organizations remained small undertakings. Isolated and uncoordinated, these fragile organizations were vulnerable to shifts in community support and perceptions of need. With health departments, visiting nurse associations (VNAs), other voluntary organizations, and private-duty nursing registries providing an uncoordinated assortment of curative and preventive nursing services, gaps and duplication were inevitable. Despite three decades of experiments, demonstrations, and studies attempting to establish comprehensive coordinated home care, little changed.

By the 1930s, home care reached a turning point. For VNAs and for most communities, the circumstances that had originally created a demand for services no longer seemed to exist. Urban death rates declined dramatically, and chronic disease replaced infection as the leading cause of death. Importantly, chronic disease did not frighten the public or stimulate philanthropic responses. The original mission of the VNAs became increasingly elusive, and support gradually declined. Simultaneously, medical, surgical, and even some obstetrical patients of all classes began to seek hospital care. With more middle- and upper-class patients using hospitals, private-duty nurses predictably followed them into these institutions. For family caregivers, the trend toward hospitalization became a partial solution to the endless obligations of caring for the sick at home. At a time when home care offered a less expensive alternative to institutional care, the home was nevertheless perceived as a less desirable locus of care. Throughout the 1940s, the ascendancy of chronic illness and the growing stature of the hospital dominated the nation's health care agenda. The work and financing of home care continued to be overshadowed and recast by these realities.

For VNAs in the 1950s, the number of home visits dwindled, costs spiraled, and deficits expanded. While many hoped that the growing availability of health insurance would relieve financial worries, home care programs struggled to survive. Amazingly, by the end of the decade, home care was being reinvented. This "back-to-the-home movement" provided the groundwork for federal policy on home care that would follow in the 1960s.

Medicare: Home Care's Second Coming

Across the United States, implementation of the Medicare program on 1 July 1966 marked the beginning of a new era for home care. Almost overnight, Medicare redefined home care as an alternative to hospital care. What resulted was a narrowly defined, fragmented, and uncoordinated set of acute-care services poorly adapted to the needs of the chronically ill at home. Although expenditures grew, many families of the elderly and their physicians found home care unfeasible and institutional care much less trouble. Policymakers and third-party payers who regarded home care of little intrinsic value continued to search for evidence of cost effectiveness. As always, the majority of care at home was provided by informal support systems—family, friends, or neighbors, mostly women.

By the 1970s, home care became part of the debate about long-term care for the elderly. This time, it was examined as a cost-effective substitute for nursing-home care. Confronted with a growing number of elderly people in need of care, a variety of research studies and federal demonstration projects examined the impact of home care on the total cost of care. After a decade of study, the findings were not encouraging. It was generally agreed that home care did not significantly reduce hospitalization or nursing-home use. It also became clear that patients did not have the same goals for home care as did payers or policymakers. They simply sought relief—as caregivers, paying consumers, and managers of care.[1]

Throughout the 1980s, a variety of legislative and judicial actions and revisions of Health Care Financing Administration policies produced expansion of home care coverage and removed disincentives for home care use. Changes in hospital reimbursement resulting in earlier hospital discharge, combined with new portable technologies, also moved services such as intravenous nutrition, chemotherapy, respiratory therapies, dialysis care, and other high-tech therapies into the home setting. Rapid growth combined with frenetic transfer of technology to the home immediately became home care's next challenge.

As in the hospital, the marvels of technology came with a price. This new technological challenge created substantial gaps between technical prowess and the humane, just, and efficient use of technology. Questions of access, equity, standards of quality, and consequences for families were quickly raised. Over time, it became clear that sending patients home to manage and monitor complicated equipment and materials that required special knowledge, skill, and composure could cause them harm.[2]

By 1994, as the result of a 22 percent growth in the over-sixty-five population in the 1980s, high-tech advancements, changing financial incentives, and consumer preference for care at home, both the number of agencies providing home care and Medicare home health care spending increased dramatically. Reluctantly, policymakers acknowledged the necessity of planning for the acute, primary, and long-term care needs of an aging population and its projected epidemic of disability and dependency. Simultaneously, rising home care use and expenditures created the demand for tighter management of the costs, quality, and outcomes of care at home. The challenge remained to create an affordable model of coordinated and integrated care.[3]

By 1996, the National Health Policy Forum concluded that it was the lack of consensus on definition and objectives that made development of public policy for home care so difficult. The forum cautioned, "Home care is at once both a formal service provided by paid staff and assistance furnished by such informal supporters as families and relatives, a medical and a nonmedical intervention, acute and long-term care, and a service with merits of its own and one whose value is assessed in terms of reductions in institutional care. These dichotomies affect the principal goals of home care, its component elements, and its place within an overall continuum of health care."[4]

This contemporary summary reflects the entire history of home care. Once again, policymakers are revisiting its costs and benefits. Providers of home care are adaptive and persistent, adjusting case mix and patterns of care to match available reimbursement and somehow offering services needed or desired by the sick and vulnerable. It is no surprise, however, that home care is not always a thrifty substitute for costly institutional care—it is simply care at home![5] The effectiveness of home care is difficult to quantify, but the large reservoir of need makes growth difficult to curtail. As one group of authors has suggested, home care may be our one way to institutionalize "caring."[6]

Reining in a Benefit Out of Control

By 1996, Medicare coverage for home care was described as a benefit out of control. The milestones in this evolution are well documented in the literature and have been examined endlessly by legislators, policymakers, and health services researchers. Regardless of its pros and cons, home care was the fastest growing component of the Medicare program during the 1990s. Representing 10 percent of all Medicare costs, home care accounted for $16 billion in spending annually.[7]

Over a thirty-year period, a program designed to meet the needs of people with short-term acute illness was also providing long-term care to the chronically ill. Ambiguity over the interpretation of benefits had created the opportunity for providers of home care to recast the Medicare regulations. Just like the Metropolitan Life Insurance Company in earlier years, the government suddenly found it could not manage or control home care. While presumed reasonable, necessary, and medically appropriate, the expansion of home care was deemed unsustainable; policymakers' demands for reform were heard once again.[8]

For policymakers faced with the potential insolvency of the Medicare program, the growth of care at home raised the all too familiar questions of how much and what kind of home care we are willing to pay for, who should receive that care, who should provide it, and for how long. A century of experimentation in home care had clearly outlined the parameters of the controversy and its likely outcome. As disenchantment with home care's assumed ability to reduce the costs of health care grew, the debate was further politicized by charges of overutilization, fraud, and abuse. Politicians and the public were reminded of the impossibility of policing what happens in the privacy of the patient's home.[9]

In this threatening climate, thousands of small, competitive, vulnerable, and disjointed home care agencies were called upon to unite. The National Association for Home Care served as the focal organization for these efforts. Founded in 1982, this organization is the largest trade association serving the interests of home care, hospice, and home health aide organizations nationally. Its chosen sound bite for the 1997 federal budget debates was "Home Care: It Works, It Saves Money, and It Keeps Families Together." As the 1997 federal budget debates raged on, however, home care's message was not marketable politically.[10]

On 1 October 1997, the Medicare home care benefit enacted in 1965 was for all practical purposes discontinued without public discussion. This time, home care's threatened demise was the result of cost-containment efforts to reduce the federal deficit and to ensure the solvency of the Medicare program. Predictably, policymakers once again failed to confront home care's peculiar, reverberating, and fundamental underlying ambiguities. Not addressed were the really tough political issues: Is caring for the sick at home a private family obligation or a responsibility shared with a caring society? Should home care be provided only under the most restrictive of circumstances, or whenever it can help? Or should we simply not decide and just keep muddling along?

The outcomes of this so-called reform were swift and dramatic: more than three thousand home care agencies closed, the number of visits per patient decreased, and public funding for home care was significantly reduced. The greatest impact was in the reduction of home health services for persons with medically complex chronic illnesses. It appeared that home care in the twenty-first century would be characterized by fewer services, more family caregiving, and new technologies such as tele–home care.[11] Confronted with shortsighted economic and political attacks, home care's proponents were once again forced to reinvent their purpose, mission, and methods.

The combination of an aging population and a decline in available family caregivers only emphasizes an expanding need for home care services currently not covered by any form of third-party coverage. At present, one-third of home care services are purchased out-of-pocket by families. Most of these services are obtained from organizations reminiscent of private-duty registries of the past, with an estimated ten thousand agencies offering some form of home care, from brief visits to twenty-four-hour live-in care.[12]

Today, a significant but largely invisible group of "informal caregivers" provides most care for the sick at home. Despite an estimated at-market value for their work of $196 billion, there is little support to sustain the efforts of these family caregivers. As in the past, this obligation falls disproportionately on women. For these women, "caring" remains inconvenient and challenging and interrupts daily patterns of living and working. In our culture, the demands of home care are grossly undervalued and codified as housework, family obligation, or perhaps a voluntary charitable responsibility. During the nearly two centuries examined in this book, the circumstances of patients and families in need of home care have changed very little—women manage on their own as best they can. In a society that has never confronted our inability to "value" women's work, it is not surprising that we have such difficulty measuring, quantifying, and paying for "home care." Instead of acknowledging the home as the preferred site of caregiving, we see it as a repository of unmet need.

Sickness and injury are not congruent with a production-oriented and worker-dependent society. Assistance from family, benevolent volunteers, or paid caregivers is generally required (although not always available) and transcends class, race, ethnicity, age, gender, place, and time. The fear of "failure to die or get well" remains a central concern in any discussion of home care; whatever the configuration of services, long-term care is a particular cause of uneasiness. How to determine the appropriate recipients and payers is a fundamental and unresolved problem, after years of inconclusive debate. The open-ended and private nature of care at home has spawned endless worry and a general unwillingness to pay for the home care of others.

Predictably, home care survived at the margins of the health care system while the hospital was transformed into the uncontested site for giving birth, receiving treatment for illness, and dying. The prominence of the hospital obscured the unseen yet steady and competent visits sustaining the sick at home.[13] While we are willing to invest in "scientific" advancements and hospital care, ours is not a caring society that routinely asks families in need how we can help.

Home Care's Future?

On 14 November 1999, the front page of the *New York Times* announced that, for the first time ever, annual spending for Medicare had dropped. This reversal in spending actually surprised many health policy experts. Given the hopes that the Balanced Budget Act of 1997 would simply slow the growth of Medicare, the *decline* in spending was declared a phenomenal development. While the *Times* acknowledged that reductions in home care contributed to this dip in spending, it failed to mention the personal consequences of this policy decision.[14]

By April 2000, *New York Times* headlines proclaimed "Medicare Spending for Care at Home Plunges by 45%." This article described how, in just one year, the number of people receiving home care shrank by six hundred thousand, making the consequences of limiting payments for home care inescapable. The facts were clear: home care agencies simply could no longer stay in business if they accepted elderly patients requiring long-term or complex care. Most disturbing was the news that some agencies now accept only those patients who have "a close relative who can provide some of the care." As government bureaucrats publicly expressed their surprise at the extent of reductions and health policy experts predicted longer and costlier stays in hospitals and nursing homes, advocates for the elderly declared alarm at reports that our most vulnerable elderly were unable to get "home care help." Politicians, hoping to be reelected and eager to allay voters' concerns, declared the cuts "far deeper and more wide-reaching than Congress ever intended" and promised a bipartisan remedy. Declaring this an incident of "good intentions . . . gone awry," a *Times* editorial suggested that it would be prudent for Congress to cancel future reductions and give serious attention to "what the payment system [for home care] ought to be." Perhaps this time an analysis of how we pay for home care should be preceded by a serious consideration of what kind of "home care help" we as a society want to purchase. Hopefully, this time we can get past our ambiguities about what our neighbors deserve and our fear that somehow we will end up paying for someone else's "housework."[15]

For those home care providers who survived the first wave of draconian cuts, proposed rules for Medicare's new home care prospective payment system promise a better future for providers and recipients of home care. As in the past, changes in financing mean yet another reinvention of care at home. While long-range planning awaits publication of the final regulations in July 2000, dramatic changes in visiting patterns, case mix, provision of therapies, and length of service are expected. Once again, the promise is for clinical needs to determine the level and amount of care received at home.

Simultaneously, the implications of a June 1999 Supreme Court ruling (*Olmstead v. L.C.*) were being declared "deep and profound" by state officials across the country. By mandating the redirecting of state spending from institutions to community-based care, the Court had created the opportunity for people of all ages with all types of disabilities to be cared for at home. Then, in February 2000, the Clinton administration directed states to evaluate hundreds of thousands of people in nursing homes, mental hospitals, and state institutions for the possibility of care at home. Finally, by declaring unnecessary institutionalization of individuals with disabilities discriminatory under the Americans with Disabilities Act, Donna Shalala, Secretary of Health and Human Services, essentially extended a vast array of home care services to those currently cared for in institutions, as well as to underserved disabled people already living at home. Compliance with these federal mandates could potentially cost more than $2 billion a year nationwide. While state officials predict that the *Olmstead* decision will "bust the bank," it is likely to lead to dramatic changes for thousands of people living with disabilities.[16]

Thus, in a few short months, the future of home-based care for persons who are acutely or chronically ill has been completely rewritten. In the final analysis, the problem is not the unwillingness of providers to change patterns of care in a changing world. More to the point, the story of caring for the sick is one in which the problems are keenly understood and the solutions do exist.[17] We certainly know how to provide and finance community-based care in this country. An outstanding example is PACE (Program of All-inclusive Care for the Elderly), a comprehensive service delivery and financing model of acute, primary, and long-term interdisciplinary care.

As this final (for now) chapter attests, the money can always be found when sufficient public will exists to change the status quo. In a recent *New England Journal of Medicine* letter to the editor, Anne Somers, noted health care expert and experienced family caregiver, described her personal experiences with the callousness of a prosperous society that turns its back on this devastating problem of caring for those with chronic illnesses. She concludes that decent care at home is entirely possible. "It is not cheap, but it is feasible. The lack of such relief is primarily a problem of public policy and political will."[18] The inability of the public to visualize the elements, outcomes, or value of home care only compounds the difficulty of deciding whether care of the sick at home is our civic duty or a private family responsibility. Perhaps it is the gendered and private nature of home care that makes commitment to this "institutionalization of caring" so difficult. On this point of divergence, we are left to this day with the unanswered question, "Who pays, for whom, for what, and how much?"

Abbreviations

AJN	*American Journal of Nursing*
AJPH	*American Journal of Public Health*
ANA	American Nurses Association
Banks Collection, S.C.	Anna DeCosta Banks Papers, Waring Historical Collection, Medical University of South Carolina, Charleston, South Carolina
Banks Collection, Va.	Anna DeCosta Banks Papers, Hampton University Archives, Hampton, Virginia
CHA	Boston's Community Health Association
CPP Collection	Directory for Nurses and Historical Collections, College of Physicians of Philadelphia, Philadelphia, Pennsylvania
DHEW	U.S. Department of Health, Education and Welfare
DNAP Collection	District Nurses Association of Providence, Rhode Island, Collection, Visiting Nurse Association of Rhode Island, Providence, Rhode Island
GCNA Collection	Greater Cleveland Nurses Association Records, Western Reserve Historical Society, Cleveland, Ohio
GNAV Collection	Graduate Nurses Association of Virginia, Virginia Nurses Association Collection, Archives and Special Collections, Tompkins-McCaw Library, Medical College of Virginia, Virginia Commonwealth University, Richmond, Virginia
IDNA Collection	Collection of the Instructive District Nursing Association of Boston, 1886–1922; Community Health Association, 1922–1942; Visiting Nurse Association thereafter; Mugar Library, Boston University, Boston, Massachusetts
JAMA	*Journal of the American Medical Association*
LBS Collection	Ladies Benevolent Society Collection, South Carolina Historical Society, Charleston, South Carolina
MLI Collection	Metropolitan Life Insurance Company, Metropolitan Life Insurance Company Library and Archives, New York

MNA Collection	Massachusetts Nurses Association Collection, Mugar Library, Boston University, Boston, Massachusetts
NOPHN Collection	National Organization for Public Health Nursing Collection, National League for Nursing, New York
NYPL	New York Public Library
Opportunity	*Opportunity: Journal of Negro Life*
PHN	*Public Health Nurse Quarterly*, 1913–1918; *Public Health Nurse*, 1919–1931; *Public Health Nursing* thereafter
Trained Nurse	*Trained Nurse and Hospital Review*
VNAC Collection	Chicago Visiting Nurse Association Collection, CareMed Chicago, Chicago, Illinois
VNAR Collection	Instructive Visiting Nurse Association of Richmond Collection, Archives and Special Collections, Tompkins-McCaw Library, Medical College of Virginia, Virginia Commonwealth University, Richmond, Virginia
VNQ	*Visiting Nurse Quarterly*
VNSNY	Visiting Nurse Service of New York
VNSNY Collection	Henry Street Settlement Collection, Visiting Nurse Service of New York, New York
VNSP Collection	Collection of the Visiting Nurse Society of Philadelphia, 1886–1959; Community Nursing Services of Philadelphia, 1959–1979; Community Home Health Services of Philadelphia, 1979–1990; Visiting Nurse Association of Greater Philadelphia thereafter; Center for the Study of the History of Nursing, School of Nursing, University of Pennsylvania, Philadelphia, Pennsylvania
Winslow Collection	C.-E. A. Winslow Collection, Sterling Memorial Library Archives, New Haven, Connecticut

Notes

Preface

1. Ruth Hubbard, "The Development of Public Health Nursing in the United States," in National League for Nursing, *Report of Conference on Public Health Nursing Care of the Sick at Home* (New York: National League for Nursing, 1953), 44.

Prelude

1. Walter Fraser, Jr., *Charleston! Charleston! The History of a Southern City* (Columbia: University of South Carolina Press, 1989), 213–46. For an excellent discussion of the "peculiar South," see Drew Gilpin Faust, "The Peculiar South Revisited: White Society, Culture, and Politics in the Antebellum Period, 1800–1869," in John Boles and Evelyn T. Nolan, eds., *Interpreting Southern History: Historiographical Essays in Honor of Sanford W. Higginbotham* (Baton Rouge: Louisiana State University Press, 1987), 78–119.

2. Fraser, *Charleston!*, 46, 186.

3. Robert Rosen, *A Short History of Charleston* (San Francisco: Lexikos, 1982), 67; Fraser, *Charleston!*, 229.

4. Fraser, *Charleston!*, 197, 217, 240.

5. Ibid., 105–6, 220, 227; William H. Pease and Jane H. Pease, *The Web of Progress: Private Values and Public Styles in Boston and Charleston, 1828–1843* (New York: Oxford University Press, 1985), 11, 44.

6. Fraser, *Charleston!*, 217; Joseph Ioor Waring, *A History of Medicine in South Carolina, 1825–1900* (Columbia, S.C.: R. L. Bryan, 1967), 40–41.

7. Waring, *History of Medicine*, 1–13, 71–85, 86–91. See also Susan Smith, *Sick and Tired of Being Sick and Tired* (Philadelphia: University of Pennsylvania Press, 1995); Todd Savitt, *Medicine and Slavery: The Diseases and Health Care of Blacks in Antebellum Virginia* (Urbana: University of Illinois Press, 1978).

8. "Management of a Southern Plantation: Rules Enforced on the Rice Estate of

P. C. J. Weston, Esq., South Carolina," *De Bow's Review* 22 (n.d.): 38–44. For a discussion of care of the sick on plantations, see Savitt, *Medicine and Slavery,* 5–13.

9. Minutes of the Board, Journal 2, 15 September 1843, LBS Collection.

10. Barbara Bellows, "Tempering the Wind: The Southern Response to Urban Poverty, 1858–1865" (Ph.D. diss., University of South Carolina, 1983), 1–38; Pease and Pease, *Web of Progress,* 115–37.

11. Waring, *History of Medicine,* 14–17; Benjamin J. Klebaner, "Public Poor Relief in Charleston, 1800–1860," *South Carolina Historical Magazine* 55 (October 1954): 210–20.

12. Between 1830 and 1848, the almshouse admitted an average of 332 poor persons (called pensioners) each year, and there were 176 outdoor pensioners. In 1824, 170 pensioners received 231 rations, some families receiving more than one ration. A ration provided four pounds of bread, four pounds of beef, and three quarts of rice or five quarts of grits per week. In 1838, pensioners also received four quarts of soup (bread, beef, and rice). In 1855, 252 rations were issued. By 1859, reflecting rising unemployment, the number had increased to 386. Klebaner, "Public Poor Relief."

13. Free blacks could own property (including slaves), marry, learn to read and write, organize churches, and engage in trade. Although their freedoms were generally protected by law, free blacks were often treated unfairly. Rosen, *Short History of Charleston,* 73; Fraser, *Charleston!,* 242; Pease and Pease, *Web of Progress,* 9; "Report on the Free Colored Poor of the City of Charleston," *Charleston Courier,* 10 December 1842.

14. Fraser, *Charleston!,* 136, 153; Waring, *History of Medicine,* 107, 126–27, 136, 140, 148.

15. The LBS claims to be the country's oldest women's organization, because the Female Benevolent Society of Wiscasset, Maine, founded earlier (in 1805), no longer exists. See Margaret Simons Middleton, "The Ladies Benevolent Society of Charleston, S.C.," *The Yearbook of the City of Charleston* (Charleston, S.C.: Walker, Evans, and Cogswell, 1941), 216–39. For a discussion of southern women's volunteerism, see Anne Firor Scott, *The Southern Lady: From Pedestal to Politics, 1830–1930* (Chicago: University of Chicago Press, 1970); and Jacquelyn D. Hall and Anne Scott, "Women in the South," in Boles and Nolan, *Interpreting Southern History,* 454–509.

16. Pease and Pease, *Web of Progress,* 138–52. Jane and William Pease have argued that Charlestonians' unusual inclusiveness stemmed from their fear that any exclusiveness was a threat to white solidarity. Bellows makes a similar argument in "Tempering the Wind," 1–38.

17. Pease and Pease, *Web of Progress,* 122, 145.

18. Jane H. Pease and William H. Pease, *Ladies, Women, and Wenches: Choice and Constraint in Antebellum Charleston and Boston* (Chapel Hill: University of North Carolina Press, 1990), 122–23.

19. Ann Mitchell, first LBS superintendent, described this religious orientation in her 1824 annual speech to the membership; see Minutes of the Board, Journal 2, 15 September 1824, LBS Collection. See also ibid., 15 September 1825, 15 September 1843, and 15 September 1847.

Charleston's benevolent societies of the 1830s and 1840s provided an estimated $25,000 annually for the sick and the needy. The LBS during this period had an annual budget of about $2,500, most of which went directly into patient care; see Pease and Pease, *Web of Progress*, 145. See also Bellows, "Tempering the Wind," 1–38.

20. The first treasurer, Sarah M. Drayton, a single woman, was succeeded in 1825 by Mrs. Parker, perhaps a widow. Of the nine treasurers between 1813 and 1896, four were listed as "Mrs." The Act of Incorporation for the LBS stipulated that the property (lands, tenements, money, goods, and chattels) held by the society should not exceed an annual value of $5,000; see Constitution, 1823, LBS Collection. Margaret Simon Middleton would later conclude in her history of the society that "the success of this Society perhaps lies in the fact that its officers do not change very often and their loyalty keeps it going through difficult times"; see Middleton, "Ladies Benevolent Society."

21. Middleton, "Ladies Benevolent Society"; Journal 2, LBS Collection. Between 1813 and 1824, the LBS cared for 2,916 patients. Analysis of visitors' reports suggests that the ladies often visited alone and that occasionally only one visitor was available to cover a whole ward. See, for example, Reports of the Visiting Committee, Journal 2, 15 September 1831 to 15 December 1831, 15 December 1831 to 15 March 1832, and 15 September 1833, LBS Collection. See also Elise Pinckney, "The LBS of Charleston: Oldest of Its Kind in the Country," *South Carolina Historical Magazine* 54 (July 1953): 5, 22–23; Journal 2, 15 September 1824 and 15 September 1833, LBS Collection.

22. Rules of the Visiting Committee were published annually in the *Charleston Courier* (the source of the quotations in the text).

23. For a brief period, the ladies tried to create rules that specified the amounts of supplies to be given, but sickness and poverty proved difficult to quantify and these rules were short lived. Constitution, 1823; Journal 2, June 1834, 15 September 1834, and 15 September 1843; all in LBS Collection.

24. Constitution, 1831; Journal 2, 15 September 1843; both in LBS Collection.

25. Journal 2, 15 September 1834, LBS Collection.

26. Annual Report 1892, LBS Collection; Middleton, "Ladies Benevolent Society," 237.

27. R. Righton, Mary Jones, and Mary Parker, "Plan Respecting Pensioners," 15 June 1825, LBS Collection. Note that *pensioners,* unlike *patients,* were given money ($0.375 weekly); see Constitution, 1852, LBS Collection.

28. Annual Report of the Board, Journal 2, 15 September 1825, LBS Collection.

29. Rules of the Visiting Committee, 1851; see also Journal 2, 15 September 1847 and 15 September 1850; all in LBS Collection.

30. Rules of the Visiting Committee, 1851; Constitution, 1831; Journal 2, 26 October 1857; all in LBS Collection.

31. Journal 2, 26 October 1857, LBS Collection.

32. According to historian Bernard Powers, because "free Negroes" of mixed blood constituted about 40 percent of all free blacks, in South Carolina they were often referred to as "free persons of color." Bernard Powers, Jr., *Black Charlestonians: A Social History, 1822–1885* (Fayetteville: University of Arkansas Press, 1994), 36, 38, 51, 52, 55.

33. Annual Report of the Board, Journal 2, 15 September 1836, LBS Collection. Powers contends that white Charlestonians had a "congenital suspicion" of free blacks. For example, the Act of 1820 required free blacks to register twice a year and explain any prolonged absence from the city. See Powers, *Black Charlestonians,* 36, 49, 62.

34. The Proceedings of the Annual Meeting, 15 September 1825 (Journal 1, p. 57, LBS Collection), reported that the "Resolution respecting free people of color was read. A motion was made and unanimously resolved. The motion was made by Mrs. Parker and seconded by Mrs. Jones, two of the three members of the Committee on Pensioners." See also Preamble, Act of Incorporation and Constitution of the LBS, 1823, p. 5, LBS Collection.

35. Minutes of the Board, 25 June 1860, LBS Collection. The board had asked the lawyer Henry Lesesne for his opinion on the meaning of the term *free persons of color*. He replied that, since the enactment of more stringent emancipation laws, it was impossible to say what was sufficient evidence of legal recognition. He concluded that payment of taxes would be enough evidence of "free" status. After 1841, loopholes in the state's regulations were corrected with passage of the Act to Prevent the Emancipation of Slaves. This law voided any bequest, trust, or conveyance for the removal of slaves out of state and bequests that made provisions for nominal servitude. See Powers, *Black Charlestonians*, 37–40.

36. Annual Report of the Board of Managers, Journal 2, 15 September 1936, LBS Collection. John M. Hopkins, born in Haddonfield, New Jersey, went into business in the South because the milder climate better agreed with his health; see Middleton, "Ladies Benevolent Society." Marie Drayton received $5 every month until she died in 1843; see Journal 2, 15 September 1843, LBS Collection.

37. Rules for Distribution of Hopkins Funds, 15 December 1835, LBS Collection.

38. Annual Report of the Board of Managers, Journal 2, 15 September 1836, 1839, 1840, 1843, and 1847; Will of John M. Hopkins, 18TS., No. 44, 301–3; 1810 Hopkins Fund Account, 1835–1930—all in LBS Collection; U.S. Census, Charleston, S.C.

39. In 1829, Kohne's husband, Frederick, left $5,000 to the LBS—its largest legacy. Elizabeth Kohne had been a lifetime member since 1817, but never an officer or a visitor. See Journal 2, 15 September 1836 and 15 December 1835, LBS Collection; Report on the Free Colored Poor (Pamphlet), 1842; Joseph W. Barnwell, "The Centennial Address," 11 February 1913, *Centennial Pamphlet* (Columbia, S.C.: R. L. Bryan, 1913), 12. Mrs. Kohne's bequest was added in a second codicil to her will "with the aid of her marriage settlement" on 1 April 1850; see "Mrs. Eliza Kohne's Will," Library Company of Philadelphia, Am 1853 Koh 14506.O.11.

40. Constitution, 1823; Annual Report 1825 and Journal 2, 15 December 1827, 15 September 1831, 15 September 1840, and 15 September 1847; all in LBS Collection.

41. From 1814 to the late 1820s, the LBS received the proceeds of an annual sermon preached by the clergy of the city and a "very handsome and free will offering by the Hebrew congregation." Between 1814 (when the first sermon was preached, at St. Michael's) and 1826, the society received fourteen collections amounting to $1,987; see Journal 2, 15 September 1826, LBS Collection; Middleton, "Ladies Benevolent Society," 216–39; and Journal 2, 15 June 1827; Annual Report 1831, 146; Journal 2, 25 October 1858; Annual Reports 1826 and 1838; all in LBS Collection.

42. See Annual Reports, LBS Collection. The permanent fund had $22,061 and the Hopkins Fund $15,154. In 1824, subscriptions contributed $1,665 and investments $464 toward a budget of $2,684.

43. Roper Hospital, a municipal hospital and teaching institution, was fully opened in

1856. For many years Charleston had a number of small private hospitals and "sick houses" for slaves; see Waring, *History of Medicine,* 17–21. The "efficiency" of the Howard Association and Roper Hospital "so liberally aided as they were in times of epidemics" made it unnecessary for the ladies to "seek patients that our means would not allow us to do much for." The Methodist Benevolent Society, Howard Association, and Young Men's Christian Association also visited the sick. See Record Book, 14 December 1839 and 25 October 1854, LBS Collection.

As early as 1825, there is mention of Roman Catholics being referred to the Sisters of Charity and later the Sisters of Mercy; both were, no doubt, the Sisters of Charity of Our Lady of Mercy. See Record Book, 15 March 1839, 27 June 1871, 14 December 1839, 15 June 1840, and 16 December 1844, LBS Collection. See also Sister Anne Francis Campbell, "Bishop England's Sisterhood, 1829–1929" (Ph.D. diss., St. Louis University, 1968).

By midcentury, the women religious in the United States numbered fifteen hundred and had established forty-one communities, whose main purpose was to establish and staff institutions of education, health, and social welfare. Between 1830 and 1840, when yellow fever and cholera were epidemic in Charleston, the sisters ran a small hospital/ house to care for the sick poor. See Sister Mary Denis Maher, *To Bind up the Wounds: Catholic Sister Nurses in the U.S. Civil War* (New York: Greenwood Press, 1989), 15, 31; Flora Bassett Hildreth, "The Howard Association of New Orleans, 1837–1878" (Ph.D. diss., University of California, Los Angeles, 1975).

Chapter 1: Trained Nurses for the Sick Poor

1. Paul Boyer, *Urban Masses and Moral Order in America* (Cambridge: Harvard University Press, 1978), 3–21, 123–90; Walter Tratter, *From Poor Law to Welfare State* (New York: Free Press, 1974); Roy Wiebe, *The Search for Order: 1877–1920* (New York: Hill and Wang, 1967).

2. Wiebe, *Search for Order;* Carol Smith-Rosenberg, *Religion and the Rise of the American City: The New York City Mission Movement, 1812–1870* (Ithaca: Cornell University Press, 1971); Michael Katz, *In the Shadow of the Poorhouse: A Social History of Welfare in America* (New York: Basic Books, 1986), 58–84.

3. Katz, *Shadow of the Poorhouse,* 58–84.

4. Eileen M. Crimmins and Gretchen Condran, "Mortality Variations in U.S. Cities in 1900," *Social Science History* 7 (1983): 31–59; Gretchen A. Condran, "Changing Patterns of Epidemic Disease in New York City," in David Rosner, ed., *Hives of Sickness: Public Health and Epidemics in New York City* (New Brunswick, N.J.: Rutgers University Press, 1995), 27–41; Judith Walzer Leavitt and Ronald Numbers, "Sickness and Health in America: An Overview," in Judith Leavitt and Ronald Numbers, eds., *Sickness and Health in America* (Madison: University of Wisconsin Press, 1997), 3–10.

5. For example, Abbie Howes, founder of the Instructive District Nursing Association of Boston (IDNA), observed the English system in 1884, corresponded with William Rathbone (founder of district nursing in Liverpool), and sent the IDNA's nurse (Miss Gordon) to study the English system in 1887. Emile M. Mansel to Abbie Howes, 12 December 1885 and 11 August 1887, IDNA Collection, BU N34, box 1, folder 1.

6. Rathbone's ideas about constructive systems of relief were essentially the same as those of the Charity Organization Movement in America; see Boyer, *Urban Masses,* 143–61. For an extensive discussion of the origins of district nursing in England, see Annie Brainard, *The Evolution of Public Health Nursing* (Philadelphia: W. B. Saunders, 1922), 103–11.

7. Brainard, *Evolution of Public Health Nursing;* William Rathbone, *Sketch of the History of District Nursing* (London: Macmillan, 1890), 15–18.

8. Rathbone, *Sketch,* 15–18.

9. Brainard, *Evolution of Public Health Nursing,* 111–13; Rathbone, *Sketch,* 19–24.

10. Committee of Home and Training School, *Organization of Nursing in a Large Town* (Liverpool: A. Holden, 1865); Florence Nightingale, "Suggestions for Improving the Nursing Service of Hospitals and on Methods of Training Nurses for the Sick Poor" (August 1874). For an excellent overview of the interactions between Rathbone and Nightingale, see Lois A. Monteiro, "Florence Nightingale on Public Health Nursing," *AJPH* 75 (February 1985): 181–86. Rathbone reportedly referred to Nightingale as his "beloved chief" in these matters. See Edward Cook, *The Life of Florence Nightingale* (London: Macmillan, 1913), 2, 125. Nightingale wrote both the introduction and the notes for the Committee of Home and Training School's *Organization of Nursing.* Nightingale's "Suggestions" is mentioned in Cook, *Life of Nightingale,* 253, 449.

11. These ideas first appeared in the *Times* (London) on 14 April 1876 and were later published as a pamphlet, which is reprinted in Luch Seymer *Selected Writings of Florence Nightingale* (New York: Macmillan, 1954), 310–18. Nightingale also wrote the introduction to Rathbone's *Sketch* (ix–xxii).

12. Brainard, *Evolution of Public Health Nursing,* 143–46; Rathbone, *Sketch,* 45–50. Florence Lee, according to Lavinia Dock, was considered the most highly trained nurse of her day. See M. Adelaide Nutting and Lavinia Dock, *A History of Nursing* (New York: G. P. Putnam's Sons, 1907), 2: 298–99.

13. Florence Lee, who had married in 1979, becoming Mrs. Dacre Craven, shared her findings and recommendations in the United States at the International Congress of Charities, Corrections and Philanthropy, which met in Chicago in 1893. Her paper is published in John S. Billing and Henry M. Hurd, eds., *Hospitals, Dispensaries, and Nursing: Papers and Discussions in the International Congress of Charities, Corrections, and Philanthropy, Section III, Chicago, June 12–17, 1893* (Baltimore: Johns Hopkins University Press, 1894), 547–54. Lee's book *A Guide to District Nurses* (London: Macmillan, 1889) greatly influenced early experiments in visiting nursing in the United States.

14. Amy Hughes, "The Rise of District Nursing in England," *Charities and the Commons* 16 (April 1906): 13–16; Brainard, *Evolution of Public Health Nursing,* 143–79.

15. This view of the sick poor as hazards to society is from Mabel Jacques, "Home Occupation in Families of Consumptives and Possible Dangers to the Public," *Transactions of the International Conference on Tuberculosis* 3 (1908): 564–69. See also Rosalind Shawe, *Notes for Visiting Nurses* (Philadelphia: P. Lakiston Son, 1893), 10.

For Nightingale's opinions on germ theory, see Charles Rosenberg, "Florence Nightingale on Contagion: The Hospital as Moral Universe," in Charles Rosenberg, ed., *Healing and History: Essays for George Rosen* (New York: Science History Publications, 1979), 116–36. For a fine analysis of the relationship between immigration and popular knowledge about the germ theory of disease, see Alan M. Kraut, *Silent Travelers: Germs, Genes, and the "Immigrant Menace"* (New York: Basic Books, 1994). See also Howard Markel, "'Knocking out the Cholera': Cholera, Class, and Quarantines in New York City, 1892," *Bulletin of the History of Medicine* 69 (fall 1995): 420–57; John Duffy, "Social Impact of Disease in the Late 19th Century," in Leavitt and Numbers, *Sickness and Health in America*, 395–402.

16. First Annual Report, 3; see also Second Annual Report, 17; and Fifth Annual Report, 9; all in VNSP Collection.

17. Harriet Fulmer, "History of Visiting Nurse Work in America," *AJN* 2 (March 1902): 412. See also Brainard, *Evolution of Public Health Nursing*, 195–249; Lillian Wald, *House on Henry Street* (New York: Henry Holt, 1915), 152–53; Mabel Jacques, *District Nursing* (New York: Macmillan, 1911), 1–12; Mary Beard, "Home Nursing," *PHN* 7 (January 1915): 44–47.

18. Lillian Wald, "Treatment of Families in Which There is Sickness," *AJN* 4 (December 1904): 427–31; Annual Report 1892, IDNA Collection; Shawe, *Notes for Visiting Nurses*, 10–18; Jacques, *District Nursing*, 32–42; Fulmer, "Visiting Nurse Work," 415–23; Ellen LaMotte, *The Tuberculosis Nurse: Her Function and Her Qualifications: A Handbook for Practical Workers in the Tuberculosis Campaign* (New York: G. P. Putnam's Sons, 1915), 11–19; Annual Report 1911, 28, DNAP Collection. Hospitals where visiting nurses were trained are listed in some annual reports; see, for example, Annual Reports 1905, 6–7, and 1906, 6–9, IDNA Collection.

19. Annual Reports 1905, 6–7, and 1906, 6–9, IDNA Collection.

20. Chicago VNA, "Rules for Nurses," *VNQ* (October 1907): 9.

21. By 1920, the criteria for success in visiting nursing remained essentially unchanged; they included a nurse's personality, teaching ability, nursing technique, and social understanding. See Josephine Goldmark, *Nursing and Nursing Education in the United States* (New York: Macmillan, 1923), 109. For a discussion of the hospital version of the womanly requirement for nursing (i.e., character), see Susan Reverby, *Ordered to Care: The Dilemma of American Nursing, 1850–1945* (Cambridge: Cambridge University Press, 1987), 49–52.

22. "Visiting Nurse—A Ministry of All the Talents," *VNQ* 3 (April 1911): 27–31; Shawe, *Notes for Visiting Nurses*, 10–16, 26–27; May Anna Tomlinson, "Compensation in Visiting Nursing," *PHN* 6 (January 1914): 77–78.

23. Mary Gardner, Annual Report 1911, 29, DNAP Collection. For information on salary, see Yssabella Waters, *Visiting Nursing in the United States* (New York: Charities Publication Committee, 1909). By 1920, annual salaries were about $1,300. See Goldmark, *Nursing and Nursing Education*, 101.

24. Louise Tattershall, "Pertinent Facts Relative to Salaries of Public Health Nurses," *PHN* 21 (November 1926): 605; "Salary Question," *PHN* 15 (September 1923): 481–83.

25. The IDNA program began as a four-month course. See Goldmark, *Nursing and Nursing Education*, 499–560; "Report of the Head Nurse," Annual Report 1906, VNSP Collection. This loss of visiting nurses due to the difficulty of the work was not a new problem—see, for example, Board of Managers Meeting, 1 June 1894, 7 June 1895, and 2 April 1897, VNSP Collection.

26. For an excellent discussion of Isabel Hampton and her influence on both visiting

nursing and the nursing profession, see Janet Wilson James, "Isabel Hampton and the Professionalization of Nursing in the 1890s," in Morris Vogel and Charles Rosenberg, eds., *The Therapeutic Revolution: Essays in the Social History of American Medicine* (Philadelphia: University of Pennsylvania Press, 1979), 201–44. See also Isabel Hampton, *Educational Standards for Nurses, with Other Addresses on Nursing Subjects* (Cleveland: E. C. Koeckert, 1907), 45–54.

27. Karen Buhler-Wilkerson, "Bringing Care to the People: Lillian Wald's Legacy to Public Health Nursing," *AJPH* 83 (December 1993): 1778–86.

28. For discussion of the visiting nurse's role in reducing hospital costs, see Fulmer, "Visiting Nurse Work"; "The Hospital Deficit," *Trained Nurse* 32 (February 1904): 115. See also Haven Emerson to Lillian Wald, 9 December 1905, Wald Collection, roll 8, NYPL. In 1887, the board of the VNSP sent a circular to local physicians and the president of Public Charities explaining how the visiting nurse could provide comfort to tuberculosis patients who could not be admitted to Philadelphia General Hospital and also enable some hospital patients to go home. The end result, they argued, would be less expensive for the city; see Board of Managers Meeting, Minutes of 21 October 1887, VNSP Collection. See also Charles E. Rosenberg, *The Care of Strangers: The Rise of America's Hospital System* (New York: Basic Books, 1987); David Rosner, *A Once Charitable Enterprise: Hospitals and Health Care in Brooklyn and New York, 1885–1915* (Cambridge: Cambridge University Press, 1982); Morris Vogel, *The Invention of the Modern Hospital, Boston 1870–1930* (Chicago: Chicago Press, 1980).

29. Mary Gardner, "Twenty-five Years Ago," *PHN* 19 (March 1937): 141–44.

30. For a more extensive analysis of this 1909 survey, see Karen Buhler-Wilkerson, "Left Carrying the Bag: Experiments in Visiting Nursing, 1877–1909," *Nursing Research* 36 (January/February 1987): 42–47. See also Waters, *Visiting Nursing;* U.S. Bureau of the Census, *Thirteenth Census of the United States, Taken in the Year 1910* (Washington, D.C.: U.S. Government Printing Office, 1913). While the number of VNAs had grown rapidly, the number of nurses employed by each organization remained small. In 1909, two-thirds of the agencies employed only one nurse, and only seventeen had staffs of ten or more nurses. As might be expected, the large agencies and their leaders dominated the development of the field. They created new and innovative programs, organized their own national organization (National Organization for Public Health Nursing) and journal, and procured insurance coverage for their patients.

31. Anne Firor Scott, "Women's Voluntary Associations: From Charity to Reform," in Kathleen D. McCarthy, ed., *Lady Bountiful Revisited: Women, Philanthropy, and Power* (New Brunswick, N.J.: Rutgers University Press, 1990), 35–54; Smith-Rosenberg, *Religion and the Rise of the American City;* Nathan Huggins, *Protestants against Poverty: Boston's Charities, 1870–1900* (Westport, Conn.: Greenwood Press, 1971); Allen F. Davis, *Spearhead for Reform: The Social Settlements and the Progressive Movement, 1890–1914* (New York: Oxford University Press, 1967).

32. Reverby, *Ordered to Care;* Ellen E. Baer, "Aspirations Unattained: The Story of the Illinois Training School's Search for University Status," *Nursing Research* 41 (January/February 1992): 44; Rosner, *Once Charitable Enterprise;* Rosenberg, *Care of Strangers,* 222.

33. Buhler-Wilkerson, "Left Carrying the Bag."

34. Karen Buhler-Wilkerson, "Public Health Nursing: In Sickness and in Health?" *AJPH* 75 (October 1985): 1155–61.

35. Haven Emerson, "Meeting the Demand for Community Health Work," *PHN* 16 (September 1924): 485–89.

36. Buhler-Wilkerson, "Public Health Nursing?"

37. Buhler-Wilkerson, "Left Carrying the Bag."

38. Ibid. Private agencies were largely in the Middle Atlantic (36%), New England (29%), and East North Central (13%) states. Most public agencies were in the Middle Atlantic (80%) states. The only region with no publicly sponsored visiting nurse programs was the East South Central.

39. Fulmer, "Visiting Nurse Work."

Chapter 2: Creating Their Own Domain

1. Mrs. J. B. Lowman, "Report on Co-operation," *VNQ* 4 (January 1912): 71–72. The Visiting Nurse Association of Cleveland, which at that time had a staff of only five nurses, had thirty board members.

2. Ibid. Attending to the patient and attending to the room were of nearly equal importance to the early visiting nurses and followed carefully the teachings of Florence Nightingale (see Chapter 1). For instruction in both activities, see the popular text followed by most nurses: Mrs. Dacre Craven, *A Guide to District Nurses* (London: Macmillan, 1889). See also Florence Nightingale, *Notes on Nursing: What It Is, and What It Is Not* (London: Harrison, 1860). For further discussion of these ideas, see Charles Rosenberg, "Florence Nightingale on Contagion: The Hospital as Moral Universe," in Charles Rosenberg, ed., *Healing and History: Essays for George Rosen* (New York: Science History Publications, 1979), 116–36.

3. Lowman, "Report on Co-operation," 72.

4. Mary Gardner, "Twenty-five Years Ago," *PHN* 29 (January 1937): 141–44.

5. Because a certain degree of expansion of VNAs was required to create the issues examined here, this chapter focuses on the larger associations—meaning, in 1909, any association employing more than ten nurses: Providence, R.I. (17 nurses), Brooklyn, N.Y. (13), Washington, D.C. (12), Baltimore (15), Boston (24), Henry Street, New York (50), Cleveland (29), and Philadelphia (19). Yssabella Waters, *Visiting Nursing in the United States* (New York: Charities Publications, 1909), 315–64.

6. Some of the agencies existing in 1901 are described in Harriet Fulmer, "History of Visiting Nurse Work in America," *AJN* 2 (March 1902): 415–23. For the number of nurses employed by each association and the number of new associations each year, see Waters, *Visiting Nursing*, 315–65. A general description of the turn-of-the-century VNA can be found in Gardner, "Twenty-five Years Ago." For "lady bountifuls," see Katharine Tucker, "The Relationship between the Board and Its Professional Staff," *PHN* 19 (June 1927): 295.

7. Annie Brainard, "The Administrative Side of Visiting Nursing," *PHN* 7 (January 1915): 57–96. For a similar discussion related to selection of a board, see Mary Gardner, *Public Health Nursing* (New York: Macmillan, 1916), 100–112. Complete listings of board members are given in the annual reports of the associations.

8. See Annie Brainard, *The Evolution of Public Health Nursing* (Philadelphia: W. B. Saunders, 1922), 203–49; C. E. M. Somerville, "District Nursing," in Isabel Hampton, ed., *Nursing of the Sick, 1893* (New York: McGraw-Hill, 1949), 119–27. Somerville's paper discusses the early management of Boston's IDNA. Providence's VNA, which had fewer nurses and managers, was organized in a similar fashion. The managers who worked directly with the nurses were often members of the committee on supervision of nurses, generally called simply the nurses' committee. See, for example, Mary Gardner, "A Successful Plan of Organization," *PHN* 5 (January 1913): 20–21. See also "Historical Notes from the Secretary's Minutes of the Nurse Committee's Meetings," DNAP Collection; Mary Aldis, "The Relationship of Directors and Nurses," *PHN* 5 (July 1913): 115–17; Minutes of the Managers Meeting, 29 October 1886 and May 1888, VNSP Collection.

9. Mary Beard, "Home Nursing," *PHN* 7 (January, 1915): 44–51.

10. This "corrective influence" was frequently mentioned in the writings of lady managers. See, for example, One of the Older Women, "A Decade of Change," *PHN* 5 (January 1913): 68; By One of Them [Mrs. John Lowman], "Concerning a Few of the Duties and Privileges of Trustees," *VNQ* 4 (April 1912): 25. For the benefits to nurses, see Beard, "Home Nursing," 45.

11. Beard, "Home Nursing." See also Tucker, "Board and Professional Staff." Mary Aldis, president of the Chicago VNA, describes nurses' views of directors as friends and the work of the "saving committee" in Aldis, "Relationship of Directors and Nurses," 117. John Lowman describes the unlikely proposition of trying to bring these two groups together on a more equal basis by appointing staff members to the board as "the commingling of diverse elements, the materially free but spiritually bound, with the materially bound and spiritually free"; see John Lowman, "Boards of Directors," *PHN* 5 (July 1913): 52.

12. Minutes of Managers Meeting, 16 January 1887, VNSP Collection. See also Linda Richards, *Reminiscences of Linda Richards: America's First Trained Nurse* (Boston: M. Barrow, 1929), 102. Miss Haydock's salary was $900 per year in 1888. See Minutes of Managers Meetings, January 1888 and May 1888, VNSP Collection.

13. "Rules for the Nurses," May 1888, and Minutes of Managers Meeting, June 1888, VNSP Collection.

14. The board's decision is given in Annual Report 1888, VNSP Collection. The only explanation in the annual report as to why nurses should live together is that the ladies wished to do as in England. The minutes state, however, that the desire to take over the whole house in which the society had its offices (and rented out rooms) had resulted when "early in the summer it became apparent that permitting the rooms connected with this office to be let to undergraduate doctors while young girls are employed as nurses could be a mistake"; see Minutes of Managers Meeting, summer 1888, VNSP Collection. For the response of nurses, see Minutes of Managers Meeting, October 1888, VNSP Collection.

15. Minutes of Managers Meeting, October 1889, and "Rules for Nurses," July 1890, VNSP Collection.

16. Dismissals are noted in Minutes of Managers Meeting, 27 May 1887, 2 July 1887, 1 December 1893, and 7 April 1916, VNSP Collection. One nurse was dismissed as unfit for the work despite her willingness and kindness of heart, another because she did non-

nursing work for patients. The board's ruling is in Minutes of Managers Meetings, 25 January 1893 and 1 February 1895, VNSP Collection. The ruling quoted in the text is from the 1895 minutes.

17. For examples of Mary Gardner's comments on dismissals, see her "Successful Plan," 20–28, and "Our Executive Officers," *PHN* 5 (July 1913): 61–68. Gardner describes her own experience as a visiting nurse in "Twenty-five Years Ago," 141. Her fictional account is in *Katharine Kent* (New York: Macmillan, 1946) (extract in text is from p. 620).

18. Brainard, *Evolution of Public Health Nursing*, 212.

19. Miss Beer's report, 28 November 1900, IDNA Collection, n. 34, box 2, folder 3.

20. Brainard, *Evolution of Public Health Nursing*, 212.

21. Annual Report 1912, 24, IDNA Collection.

22. This is how the Philadelphia managers described their new superintendent. Annual Report 1915, 5, VNSP Collection.

23. Annual Report 1912, 16, IDNA Collection.

24. One of the Older Women, "Decade of Change," 67, 68.

25. It is not surprising that many of the early leaders in this field were Johns Hopkins graduates. As Janet James has described, the university's reputation attracted a very different kind of woman to nursing, one with a strong educational background and prominent social position. The early Hopkins students came from middle- to upper-middle-class families, and few were dependent on their salaries for a livelihood. In addition, there was ample encouragement for Hopkins students to become visiting nurses. See Janet James, "Isabel Hampton and the Professionalization of Nursing in the 1890s," in Morris Vogel and Charles Rosenberg, eds., *The Therapeutic Revolution: Essays in the Social History of American Medicine* (Philadelphia: University of Pennsylvania Press, 1979), 213–14, 226–27. Other members of this generation of visiting nurse leaders who graduated from Hopkins included Katharine Olmstead, Ellen LaMotte, Ada Carr, Yssabella Waters, Florence Patterson, and Anne Stevens.

26. Dr. Hugh Cabot to Miss Elizabeth Cordner, 25 February 1908, IDNA Collection, n. 34, box 1, folder 2, pp. 1–2.

27. Ibid., pp. 3–5.

28. Annual Reports 1908 and 1911, IDNA Collection. Unfortunately, as the IDNA's expenses increased, income decreased. In 1911, the board reported that during the past five years expenses had increased by $22,000 while income had decreased by $5,000. That year the association reported a deficit of $5,784. See Annual Report 1911, 19, IDNA Collection.

29. "Conference on District Nursing Problems, 2/16/11," IDNA Collection, n. 34, box 2, folder 322. At the time, the IDNA had forty-three staff nurses: ten maternity nurses, eleven Boston dispensary nurses, one nursery nurse, one industrial nurse, thirteen nurses working with private doctors, six MLI nurses, and one contagion nurse. The board's conference on 15 February 1911 was convened to discuss these problems and plan for the future, and the conclusions were published in "Conference on District Nursing Problems," what the board called a "descriptive pamphlet." See also "An Address, 11/11/11," 6, and Annual Report 1912, 22, IDNA Collection, n. 34, box 2, folder 322.

30. Michael Davis to Mrs. E. A. Codman, 27 April 1911, IDNA Collection, n. 34, box 1, folder 3.

31. The IDNA board members decided that some type of "organic union" of the various unrelated groups of nurses should be created. First they would combine the different specialized groups of IDNA nurses, and by summer they hoped to unify the IDNA and the Baby Hygiene Association, an outcome that in fact required another ten years. "Conference on District Nursing Problems" and Annual Report 1911, 9, IDNA Collection.

32. Ella Crandall to Mrs. Codman, 9 November 1911; Ida Cannon to Mrs. Codman, 15 February 1911; Edna Foley to Mrs. Codman, 28 June 1911; Ella Crandall to Mrs. Codman, 25 August 1911; and Ellen LaMotte to Mrs. Codman, 26 September 1911 and 3 October 1911; all in IDNA Collection, n. 34, box 1, folder 3.

33. Elizabeth King Elliot to Mrs. Codman, n.d. [probably summer 1911]; Ella Crandall to Mrs. Codman, 31 August 1911; Mary Goodville to Mrs. Codman and Gertrude Peabody to Mrs. Codman, n.d.; and Mary Lent to Mrs. Codman, 14 September 1911; all in IDNA Collection, n. 34, box 1, folder 3.

34. Ella Crandall to Mrs. Codman, 9 November 1911, IDNA Collection, n. 34, box 1, folder 3. The board did not immediately write to Mary Beard but instead wrote Ella Crandall a second letter requesting more information about Beard. See Crandall to Mrs. Codman, 21 November 1911, IDNA Collection, n. 34, box 1, folder 3.

35. John Lewis to Mrs. Codman, 24 March 1911, IDNA Collection, n. 34, box 1, folder 3. Irene Sutliffe, who had been superintendent of New York Hospital when Mary Beard was a student there, wrote to the board that Beard had a credible record and was a well-bred, intelligent woman with "a thoroughly social point of view." She did not think Beard had "great ability" but thought she would do well as a head worker in a nursing association, "as she is in many ways fitted for that kind of work." Irene Sutliffe to Mrs. Codman, 21 December 1911, IDNA Collection, n. 34, box 1, folder 3.

36. The Annual Report of 1912 (pp. 16, 23, IDNA Collection) discusses the status of the organization. Beard's comments are in her "Report of the Work during June, 1912," IDNA Collection.

37. Annual Report 1912, 14–24, IDNA Collection.

38. Annual Reports 1914, 8, and 1915, 18, 24, IDNA Collection.

39. Annual Report 1912, 25–27, IDNA Collection.

40. Annual Reports 1913, 22; 1914, 10; 1915, 16; all in IDNA Collection.

41. Annual Report 1913, 19, IDNA Collection.

42. Ibid., 20–21.

43. Ibid., 20; "The Atlantic City Meeting (Editorial)," *PHN* 5 (July 1913): 5–12; Florence Patterson, "The First Annual Meeting of the National Organization for Public Health Nursing," *PHN* 5 (July 1913): 13–19. Louise Fitzpatrick discusses the controversy over abandoning specialization in favor of a generalized service in *The National Organization for Public Health Nursing, 1912–1952: Development of a Practice Field* (New York: National League for Nursing, 1975), 34. Mary Beard's views were first published in 1913 in her "Generalization in Public Health Nursing," *PHN* 5 (October 1913): 42–47. This debate continued well into the 1920s and was addressed by the Goldmark Report and the East Harlem Nursing Project: Josephine Goldmark, *Nursing and Nursing Education in the United States* (New York: Macmillan, 1923); East Harlem Nursing and Health Service, *A Comparative Study of Generalized and Specialized Health Services* (New York: East Harlem Nursing and Health Service, 1926).

44. Annual Report 1913, 20, 23; Annual Report 1915, 26; Treasurer's Reports, Annual Reports 1914, 1915, 1916; all in IDNA Collection.

45. Annual Reports 1911 and 1915, IDNA Collection. The additional $17,000 in wages represents the increased cost for all nurses, while increased cost for staff nurses alone was $13,914; see Treasurer's Reports, Annual Reports 1911 and 1915. See also Annual Report 1912, 20, and Treasurers' Reports, Annual Reports 1911–1915; all in IDNA Collection.

46. Supervisors Meeting, 23 November 1917, IDNA Collection. Katharine Tucker describes the transformation in the relationship between managers and nurses as a second stage that proved unsound in both theory and practice but was a necessary step before "a partnership of equals" free of patronage in either direction could be created. Tucker, "Board and Professional Staff."

47. Susan Reverby, *Ordered to Care: The Dilemma of American Nursing, 1850–1945* (Cambridge: Cambridge University Press, 1987), 143–58.

Chapter 3: "Treatment of Families in Which There Is Sickness"

The chapter title is from Lillian Wald's report, "Treatment of Families in Which There Is Sickness," *AJN*, 4 (December 1904): 427–31.

1. Alan M. Kraut, *Silent Travelers: Germs, Genes, and the "Immigrant Menace"* (New York: Basic Books, 1994), 136–65, 197–225.

2. Lillian Wald, *House on Henry Street* (New York: Henry Holt, 1915).

3. Mrs. W. T. Sedgwick, "Instructive District Nursing," *Forum* (November 1896): 297–307.

4. During the IDNA's fourth and fifth years, the Boston nurses made 40 to 47 percent of their visits with the dispensary physicians. See Annual Reports 1889 and 1890, IDNA Collection. Chicago visiting nurses also relied on pharmacies as their headquarters, publishing in the Chicago VNA's annual report lists of the pharmacies used; see Annual Report 1906, 39, VNAC Collection. See also Audrey Davis, "With Love and Money: Visiting Nursing in Buffalo, New York, 1885–1915," *New York History* (January 1990): 45–67.

5. Annual Reports 1891 and 1893, IDNA Collection.

6. Annual Report 1910, IDNA Collection; Ruth Farrisey, "Boston Visiting Nurse Association History" (unpublished manuscript, 1985).

7. Farrisey, "Boston Visiting Nurse," 15; see also ibid., 5, 8, 11, 20, 25; "Report for East Boston," 1900, IDNA Collection.

8. See, for example, Annual Reports 1910 and 1911, IDNA Collection. At that time the hospitals included Children's, Boston Floating, and Boston Lying-In. Later Mt. Sinai would establish a relationship. The IDNA also had formal relationships with the Society for Relief and Control of Tuberculosis and the Tremont Dispensary, as well as with a growing number of private physicians. In 1911, 41 percent of IDNA patients were referred from Boston Dispensary, 23 percent from Boston Lying-In, 9 percent from the MLI, and 17 percent from private doctors.

9. Report of the Head Nurse, 1906, VNSP Collection.

10. In 1895, the VNSP reported "serving" 302 physicians. Annual Reports 1895 and 1904, VNSP Collection.

11. Farrisey, "Boston Visiting Nurse," 13.

12. Wald, *House on Henry Street,* 9, 26–27.

13. See, for example, Annual Report 1918, 18, VNSP Collection. Patients were reported by families (22%), social agencies (15%), physicians (29%), the MLI (27%), and nurses (6%). In 1923, the IDNA responded to calls from six hundred physicians. Annual Report 1923, 33, IDNA Collection.

14. Mary Gardner, *Katharine Kent* (New York: Macmillan, 1946), 66–67.

15. "Report on Clinics and Health Associations," *JAMA,* 28 July 1928.

16. NOPHN, *Manual of Public Health Nursing* (New York: Macmillan, 1927), v, 8–9, 25–28. The rule requiring a physician to be in attendance is found in almost every annual report. See, for example, 26th Annual Report, 1911, 8, IDNA Collection; "Rules for the Nurses," May 1888, VNSP Collection. The Boston nurses carried their manual in their bags. See Annual Report 1924, 14, IDNA Collection.

17. Annual Report 1924, 8–9, IDNA Collection.

18. Gardner, *Katharine Kent;* Ellen LaMotte, *The Tuberculosis Nurse: Her Functions and Her Qualifications: A Handbook for Practical Workers in the Tuberculosis Campaign* (New York: G. P. Putnam's Sons, 1915), 70–89, 102–5.

19. Kraut, *Silent Travelers.*

20. LaMotte, *Tuberculosis Nurse.*

21. Most of what we know of the work of these early visiting nurses is based on how-to-do-it texts and VNAs' annual reports. In these sources, the nurses' interventions seem similar from place to place. See "Rules for the Nurses of the Visiting Nurse Association," Fifth Annual Report, 1895, VNAC Collection; Annual Report 1919, 24–25, IDNA Collection.

22. Annual Report 1905, 25, IDNA Collection; letter to Jamestown Visiting Nurse Association from Stanley Supply Co., New York, 16 May 1922, Jamestown, New York, Visiting Nurse Association Collection; Annual Report 1908, 1, VNA of Cleveland; Elizabeth Robinson Scovill, "Notes from Medical Press," *AJN* 12 (1911–12): 583.

23. "Their Health is Your Health" (fundraising booklet published by the Henry Street Settlement in 1934).

24. Nurses' Notes, "Work at Large," 21 May 1906, IDNA Collection; C. E. M. Somerville, "District Nursing," in John S. Billings and Henry M. Hurd, eds., *Hospitals, Dispensaries, and Nursing: Papers and Discussions in the International Congress of Charities, Corrections, and Philanthropy, Section III, Chicago, June 12–17, 1893* (Baltimore: Johns Hopkins University Press, 1894), 541; Rosalind Shawe, *Notes for Visiting Nurses* (Philadelphia: P. Blakiston, 1893), 80–100. Nurses' stories describing the nurse's work can be found in all the early editions of *VNQ.*

25. Harriet Fulmer, "Report of the Nurses' Work for the Year 1902," 13th Annual Report, 1903, 17, and Annual Report 1905, 38, VNAC Collection; Annual Report 1893, 8, VNSP Collection; Annual Reports 1911, 29, and 1912, 30, IDNA Collection. Boston discontinued housekeeping assistance in 1915, claiming it was no longer affordable; see Davis, "Love and Money."

26. Wald, *House on Henry Street,* 28.

27. Shawe, *Notes for Visiting Nurses,* 28–29.

28. Ibid.; Wald, "Treatment of Families"; Annual Report 1902, 17, IDNA Collection; Annual Reports 1891, 1910, and 1911, VNAC Collection.

29. The heading for this section is from John Dill Robertson, "Who Shall Nurse the Sick?" *AJPH* 11 (January 1921): 108–12.

30. Annual Reports 1909, 6, and 1913, 11, IDNA Collection.

31. "Rules for the Nurses of the VNA," Fifth Annual Report, 1895, 8–10, VNAC Collection; see also Annual Report 1895, 9, IDNA Collection.

32. For the Chicago VNA's practices, see Paula Holmes Gray, "Secretary's Report," Ninth Annual Report, 1899, 17; "Rules for Women Who Do Emergency Work for the Visiting Nurse Association"; and Harriet Fulmer's "Report of Nurses' Work," Tenth Annual Report, 1900, 13, 23; "Superintendent's Report," 11th Annual Report, 1901, 23; and "Report of Nurses' Work," 12th Annual Report, 1902, 16; all in VNAC Collection. See also Eliza Moore, "Visiting Nursing," *AJN* 1 (October 1900): 17; Annual Report 1891, 5, VNSP Collection; Annual Report 1895, 9, IDNA Collection.

As late as 1902, a variety of sources were used for emergency work in Boston. Of the 155 applicants, thirty-two were referred for care by Miss Strong's attendants, eighty-three were cared for by special nurses, and forty were cared for by the IDNA district nurses. See Annual Report 1901, 18, IDNA Collection. The MLI paid for 403 of the 455 special day and night nurses provided to Boston patients in 1910. See Annual Report 1910, 14, IDNA Collection. See also Davis, "Love and Money"; Annual Report 1917, 6, VNSP Collection; *Manual,* 19–20, NOPHN Collection. The MLI provided a brief source of funding for emergency nurses for the first two years of its nursing service. Annual reports show MLI patients receiving most of this care, and then no care.

33. Annual Report 1893 and Committee on Nursing and Physicians, 16 September and 11 November 1898, IDNA Collection.

34. Anne Goodrich, "The Need for Orientation," *AJN* 13 (February 1913): 341.

35. Isabel Lowman, "The Need of a Standard for Visiting Nursing," *VNQ* 4 (January 1912): 8–16; Davis, "Love and Money"; Grace Allen, "Shall Attendants Be Trained and Registered?" *AJN* 12 (August 1912): 933; Edna Foley, "Cornering the Employment of Practical Nurses by Visiting Nurse Associations," *AJN* 12 (January 1912): 328. For a more extensive discussion of the organization of public health nursing, see Karen Buhler-Wilkerson, "False Dawn: The Rise and Decline of Public Health Nursing in America, 1900–1930," in Ellen Lagemann, ed., *Nursing History: New Perspectives, New Possibilities* (New York: Teachers College Press, Columbia University), 89–106.

36. Wald, *House on Henry Street,* 42; Wald, "Treatment of Families"; Annual Reports 1893, 1902, 1903, and 1904, IDNA Collection; Annual Report 1893, 6, VNSP Collection.

37. Wald, "Treatment of Families"; Lillian Wald to Jacob Schiff and Mrs. Solomon Loeb, 2–7 July 1893, 2 October 1893, 3 November 1893, 4 January 1894, 2 February 1894, 1 October 1894, 8 May 1894, 8 December 1894, 11 February 1895, and 10 April 1895, Wald Collection, NYPL.

38. "Report," 1892, VNSP Collection.

39. Ibid.; Wald, "Treatment of Families," 431; Shawe, *Notes for Visiting Nurses,* 28–29.

40. Annual Reports 1902, 13; 1903, 15; and 1905, 17; all in IDNA Collection; Annual Reports 1899, 1904, 1905, and 1911, VNSP Collection; Annual Report 1906, 21, DNAP Collection.

41. For more extensive analysis of the categories of patients served, see Karen Buhler-Wilkerson, "Left Carrying the Bag: Experiments in Visiting Nursing, 1877–1900," *Nursing Research* 36 (January/February 1987): 42–47.

42. Lavinia L. Dock, "An Experiment in Contagious Nursing," *AJN* 3 (September 1903): 927–33; Katharine Tucker, "The Management of Communicable Disease from the Standpoint of a Public Health Nurse," *AJN* 22 (March 1922): 412–26; Mabel Jacques, "The Visiting Nurse in Tuberculosis: Her Importance as an Educational Agent," *Journal of Outdoor Life* 6 (May 1909): 134–37.

Technically, the term *infectious* denotes a disease that might be spread from person to person without actual contact; *contagious* disease is directly transmitted from person to person; *communicable* is a more general term that covers both infectious and contagious diseases. For an interesting note on this terminology and germs, see Nancy Tomes, *The Gospel of Germs: Men, Women, and the Microbe in American Life* (Cambridge: Harvard University Press, 1998).

43. For a discussion of tuberculosis as a diathetic disease, see Georgina D. Feldberg, *Disease and Class: Tuberculosis and the Shaping of Modern North American Society* (New Brunswick, N.J.: Rutgers University Press, 1995); and Richard Shryock, *National Tuberculosis Association, 1904–1954: A Study of the Voluntary Health Movement in the United States* (New York: National Tuberculosis Association, 1957), 28–64.

44. This view of the sick poor as hazards to society can be seen in Mabel Jacques, "Home Occupation in Families of Consumptives and Possible Dangers to the Public," *Transactions of the International Conference on Tuberculosis* 3 (1908): 564–69; Wald, "Treatment of Families"; Harriet Fulmer, "History of Visiting Nurse Work in America," *AJN* 2 (March 1909): 415–23; Mary Beard, "Home Nursing," *PHN* 7 (January 1915): 44–47; Annie Brainard, *The Evolution of Public Health Nursing* (Philadelphia: W. B. Saunders, 1922), 211; and Wald, *House on Henry Street*, 152–53.

45. Elizabeth Fee and Evelynn M. Hammonds, "Science, Politics, and the Art of Persuasion: Promoting the New Scientific Medicine in New York City," in David Rosner, ed., *Hives of Sickness: Public Health and Epidemics in New York City* (New Brunswick, N.J.: Rutgers University Press, 1995), 155–96. For an excellent discussion of the complexities of the issues related to contagious diseases, see Judith Leavitt, *Typhoid Mary: Captive to the Public's Health* (Boston: Beacon Press, 1996); and Tomes, *Gospel of Germs*.

46. According to Lavinia Dock, in 1901 there were only enough hospital beds for 7 percent of patients with contagious disease in New York City; see Dock, "Experiment in Contagious Nursing." See also Annual Report 1909, 8–9, IDNA Collection; VNA South Bend, Annual Report 1926, 4.

47. Annual Report 1909, 8–9; Secretary's Note Book, 18 March 1886; and Annual Reports 1893, 1894, 1895; all in IDNA Collection.

48. Annual Reports 1893, 1894, 1895, IDNA Collection.

49. Marguerite Wales, *The Public Health Nurse in Action* (New York: Macmillan, 1941), 159–63; Kraut, *Silent Travelers*, 105–35, 241.

50. Wales, *Public Health Nurse in Action*, 159–63; Third Annual Report, 1888, 6, VNSP Collection. In 1906, Philadelphia was interested in acquiring a nurse for working on contagious disease cases; see "Report of the Head Nurse," 1906, VNSP Collection.

51. "Rules for the Care of Infectious Cases," Fifth Annual Report, 1895, 11, VNAC Collection.

52. In 1909, only eighty-five agencies cared for patients with contagious diseases,

which included 24 percent of VNAs and 9 percent of boards of health. Yssabella Waters, *Visiting Nursing in the United States* (New York: Charities Publications, 1909).

53. Board of Managers Meeting, 31 March 1897, VNSP Collection. The number of "Americans" continued to grow. Annual reports reassured supporters that the shifts in caseload were explained by the increased numbers of babies of foreign mothers. The heading of this section on care of immigrants is from Harriet Leet, "The Problem of Many Tongues," *VNQ* 2 (July 1910): 30–39.

54. Howard Markel, "'Knocking out Cholera': Cholera, Class, and Quarantines in New York City, 1892," *Bulletin of the History of Medicine* 69 (fall 1995): 420–57; Kraut, *Silent Travelers;* Tomes, *Gospel of Germs;* Buhler-Wilkerson, "Left Carrying the Bag."

55. "Editorial Comment," *VNQ* 2 (April 1909): 7–9; Kraut, *Silent Travelers,* 197–225.

56. Annual Report 1910, 26, VNAC Collection; Nurses' Report, VNA of Dayton, Ohio, 1902; Annual Report 1909, 14–15, IDNA Collection; Annual Reports 1899, 7; 1904, 11; and 1905, 12—all in VNSP Collection; Leet, "Problem of Many Tongues"; "The New District Nurse," *PHN* 5 (January 1913): 48–53; Helen Hempstead, "The Little Submerged Patients," *VNQ* 2 (July 1910): 41–45; Harriet Mullony, "Less Familiar Friends from Central Europe," *PHN* 5 (July 1913): 99–114.

57. Nurses' Report, VNA of Dayton; "The Nurse's Story," Annual Report 1905, 14–15, and Report of the Superintendent, Annual Report 1910, 24–26, DNAP Collection. For nurses' views on various nationalities, see Amy Potts, "Visiting Nursing in Philadelphia," *Trained Nurse* 29 (August 1902): 85–88. See also "Editorial Comment," *VNQ* 1 (April 1909): 7–9; "Quarterly Report of Visiting Nurse Association," *VNQ* 1 (April 1909): 52–53; Joseph Mayper, "The Immigrant," *PHN* 5 (July 1913): 89–98.

58. See, for example, Annual Report 1892, 7–8, VNSP Collection; Davis, "Love and Money."

59. Leet, "Problem of Many Tongues."

60. Annual Report of the Visiting Nurse Association of Cleveland, *VNQ* (1912): 43; Michael Davis, Jr., *Immigrant Health and the Community* (New York: Harper and Brothers, 1921), 291.

61. The IDNA served 1,156 Italian patients (10% of the caseload). "Jewish" was listed as a nationality in the IDNA annual reports. See Annual Reports 1906 and 1913, IDNA Collection. Cleveland used the same technique—asking for support for the Jewish nurse. See Annual Report 1905, 49, VNA of Cleveland.

62. Annual Report 1923, 17, VNAC Collection.

63. Josephine Goldmark, *Nursing and Nursing Education in the United States* (New York: Macmillan, 1923), 126–27. See also Davis, "Love and Money."

64. Wald, "Treatment of Families."

65. John Farrell, "The Trend of Preventive Medicine in the United States," *JAMA* 81 (September 1923): 1063–69; Lee Frankel, "Science and Public Health," *AJPH* 5 (April 1915): 281–89; Haven Emerson, "Meeting the Demand for Community Health Work," *PHN* 16 (September 1924): 487; C.-E. A. Winslow, *The Evolution and Significance of the Modern Public Health Campaign* (New Haven: Yale University Press, 1935), 49–65.

66. Winslow, *Evolution and Significance,* 49–65. See also William Field, "Civic Control of Public Health Nursing," *PHN* 6 (October 1914): 70–80.

67. Field, "Civic Control." See also Charles Chapin, "The Evolution of Preventive Medicine," *JAMA* 76 (January 1921): 215–22. For a discussion of the changing emphasis from community to household efforts, see Barbara Rosenkrantz, *Public Health and the State* (Cambridge: Harvard University Press, 1972); and Barbara Rosenkrantz, "Cart before the Horse: Theory, Practice, and Professional Image in American Public Health, 1870–1920," *Journal of History and Allied Sciences* 29 (January 1974): 55–73. The source most often mentioned for the theory and practice of the new public health was Hibbert Hill, *The New Public Health* (New York: Macmillan, 1916). See also "Hygiene and the Nurse," *Pacific Coast Journal of Nursing* 10 (January 1914): 157.

68. Winslow, *Evolution and Significance,* 52–55. See also Charles Eliot, "The Main Points of Attack in the Campaign for Public Health," *AJPH* 5 (July 1915): 619–25.

69. The fashionable expression of this idea was the metaphor of seed and soil; see "Hygiene and the Nurse."

70. For example, in the case of the tuberculosis patient, the objective of the nurse was not so much to cure disease but to prevent its spread to individuals not yet infected. LaMotte, *Tuberculosis Nurse,* 117; "Hygiene and the Nurse"; Sir George Newman, "Preventive Medicine," *PHN* 12 (February 1920): 129–43.

71. "Hygiene and the Nurse"; M. Adelaide Nutting, "The Home and Its Relationship to Prevention of Disease," *AJN* 4 (September 1904): 913–24; Frankel, "Science and Public Health," 281–89; C.-E. A. Winslow, "Untilled Field of Public Health," *Science* 51 (January 1920): 23–33; Eliot, "Main Points of Attack," 619–25; Charlotte Aikens, "Educational Opportunities of the Visiting Nurse in the Prevention of Disease," *Proceedings of the National Conference of Charities and Corrections* (1906): 185–95; Charles Rosenberg, "Florence Nightingale on Contagion: The Hospital as Moral Universe," in Charles Rosenberg, ed., *Healing and History: Essays for George Rosen* (New York: Science History Publications, 1979), 116–36.

72. Nutting, "Home," 117–18, 918–19.

73. Ibid.; LaMotte, *Tuberculosis Nurse,* 145–48.

74. Nutting, "Home," 917.

75. LaMotte, *Tuberculosis Nurse,* 18–19. See also Annie Brainard, "The Visiting Nurse and Preventive Work," *VNQ* 1 (April 1909): 11.

76. Isabel Lowman, "A Morning with a Maternity Nurse," *VNQ* 1 (April 1909): 16–17. See also Aikens, "Educational Opportunities," 186; LaMotte, *Tuberculosis Nurse,* 59, 218–23; Mary Lent, "The True Function of the Tuberculosis Nurse," *Transactions of the International Conference on Tuberculosis* 6(1908): 78; Lillian Wald, "Educational Value and Social Significance of the Trained Nurse in the Tuberculosis Campaign," *Transactions of the International Conference on Tuberculosis* 3 (1908): 632–38.

77. Dock, "Experiment in Contagious Nursing."

78. For a more extensive discussion of Wald's views, see Karen Buhler-Wilkerson, "Bring Care to the People: Lillian Wald's Legacy to Public Health Nursing," *AJPH* 83 (December 1993): 1778–86.

79. This comment was made by Elizabeth Fox, Director of the Bureau of Public Health Nursing, American Red Cross, in Goldmark, *Nursing and Nursing Education,* 138.

80. Goldmark, *Nursing and Nursing Education,* 138.

81. Aikens, "Educational Opportunities," 186–89.

82. Ellen LaMotte, "The Unteachable Consumptive," *Transactions of the International Conference on Tuberculosis* 3 (1908): 256–60. See also LaMotte, *Tuberculosis Nurse,* 2–3, 218–23.

83. The nurses' study is reported in Winifred M. Allen and Elizabeth McConnell, "The Teachableness of the Consumptive Patient," *AJN* 15 (October 1914): 25–30. For an excellent discussion of the teachability of patients, see Barbara Bates, *Bargaining for Life: A Social History of Tuberculosis, 1876–1938* (Philadelphia: University of Pennsylvania Press, 1992), 231–40. See also LaMotte, "Unteachable Consumptive"; Lent, "True Function."

84. Lent, "True Function," 578.

85. Mary Lent, "Report of the Committee on Visiting Nursing," *AJN* 10 (June 1910): 866–68.

86. Ibid.

87. Wald, "Educational Value," 637.

Chapter 4: Caring in Its Proper Place

1. The intimate realities of caring for the sick at home suggest that caring "in its proper place" was not a purely southern conviction; even in the South, what constituted this "stiff-sided box" varied from place to place. Glenda Elizabeth Gilmore, *Gender and Jim Crow: Women and the Politics of White Supremacy in North Carolina, 1896–1920* (Chapel Hill: University of North Carolina Press, 1996), 3.

2. For a fine analysis of the intertwined relationship among race, class, and gender, see Evelyn Brooks Higginbotham, "African-American Women's History and Metalanguage," *Signs: Journal of Women in Culture and Society* 17 (winter 1992): 251–73.

3. Joseph Ioor Waring, *A History of Medicine in South Carolina, 1825–1900* (Columbia, S.C.: R. L. Bryan, 1967), 59–69, 161–71; Walter J. Fraser, *Charleston! Charleston! The History of a Southern City* (Columbia: University of South Carolina Press, 1989), 275.

4. For a discussion of scientific charity and the Charity Organization Society, see Michael Katz, *In the Shadow of the Poorhouse: A Social History of Welfare in America* (New York: Basic Books, 1986), 58–84.

5. Laylon Jordan, "The Method of Modern Charity: The Associated Charities Society of Charleston, 1888–1920," *South Carolina Historical Magazine* 88 (January 1987): 34–47; Fraser, *Charleston!,* 274–301.

6. See *Year Book: City of Charleston, S.C.* (Charleston, S.C.: Walker, Evans, and Cogswell) for the years 1880 to 1889 and 1921. See also Ruth Chamberlin, *The School of Nursing of the Medical College of South Carolina, Its Story* (Columbia, S.C.: R. L Bryan, 1970).

7. Darlene Clark Hine, *Black Women in White: Racial Conflict and Cooperation in the Nursing Profession, 1890–1950* (Bloomington: Indiana University Press, 1989), 15, 50, 57–58. See also Anna DeCosta Banks, "The Work of a Small Hospital and Training School in the South," *Eighth Annual Report of the Hampton Training School for Nurses and Dixie Hospital* (Hampton, Va., 1898–1899), 23–28. Besides the black community's obvious need for health care, without their own hospital the city's black physicians found it difficult to

survive economically. See Howard Rabinowitz, *Race Relations in the Urban South, 1865–1890* (New York: Oxford University Press, 1978), 128–51.

8. Rabinowitz, *Race Relations.* See also Vanessa Northington Gamble, "The Negro Hospital Renaissance: The Black Hospital Movement, 1920–1945," in Diane E. Long and Janet Golden, eds., *The American General Hospital: Communities and Social Contexts* (Ithaca: Cornell University Press, 1989), 20–22, 50, 82–105; Cynthia Neverdon-Morton, *Afro-American Women of the South and the Advancement of the Race, 1895–1925* (Knoxville: University of Tennessee Press, 1989); Gerda Lerner, "Early Community Work of Black Club Women," *Journal of Negro History* 59 (1974): 158–67.

9. Annual Report, 10 January 1889, and "The Ladies Benevolent Society," *Sunday News,* 1903 (newspaper clipping), LBS Collection. The LBS published its annual report in the local papers and collected these clippings in scrapbooks, which were not always carefully labeled. The notes citing newspaper clippings for this section reflect all available information.

10. Catherine Ravenel, Annual Report, 8 February 1925, LBS Collection. See also Report of the Annual Meeting, 11 January 1900; newspaper article, 1895; Record Book No. 3, Minutes of Annual Meeting, 16 January 1895; "Society Holds 115th Meeting," January 1928; Annual Report Book, 1906–1930, 92; all in LBS Collection.

11. Harriet Fulmer, "The History of Visiting Nurse Work in America," *AJN* 2 (March 1902): 415–23. Fulmer's paper was first presented at the Congress of Nurses in Buffalo, New York, 1901. See also "The 90th Anniversary of the Ladies Benevolent Society" (newspaper clipping), 1903, LBS Collection.

12. Ravenel's comments appear in several reports. See Annual Report, "Ladies Benevolent Society" (newspaper clipping), 1903; Report of the Nurses Committee, 21 January 1903; Annual Meeting, 15 January 1913; "The 89th Anniversary of the Ladies Benevolent Society" (newspaper clipping); Annual Report, *Keystone,* 21 January 1902 (newspaper clipping); all in LBS Collection.

13. "Ladies Benevolent Society," 1903; "Doers of Good Deeds" (newspaper clipping), LBS Collection; "90th Anniversary," 1903.

14. Annual Reports 1909–1911 and 21 January 1925, LBS Collection.

15. Ravenel's comments are in "90th Anniversary," 1903. See also Reports of the Nurse Committee, 21 January 1903 through 9 January 1908, LBS Collection. "She has done what she could" (describing Banks) in the heading of this section is from Anna Banks to Catherine Ravenel, Minutes of the Society, Record Book, 1906–1911, LBS Collection.

16. The description of Banks is in Annual Report 1907, LBS Collection. See also Report of the Nurse Committee, 16 June 1903, January 1907, and 5 June 1907, LBS Collection.

17. Anna Banks to Miss Sherman, 13 April 1912, Banks Collection, Va. Banks described Sherman as her teacher and well-wisher. In 1914, Banks, in addition to her visiting nurse work, was "back at the hospital again." See letters from Anna Banks to Miss Davis, 14 May 1904; to Dr. Fussell, 10 January 1900; to Dr. H. B. Fussell, 30 April 1917; to Miss Sherman, 18 April 1914; all in Banks Collection, Va.

18. Hine, *Black Women in White,* preface. The relationship between black nurses and the black community could be very complex. See, for example, Susan M. Reverby, "Re-

thinking the Tuskegee Syphilis Study: Nurse Rivers, Silence, and the Meaning of Treatment," *Nursing History Review* 7 (1999): 3–28.

19. Banks was associated with the Hospital and Training School for thirty-two years and the Ladies Benevolent Society for twenty-four years; see "The Hospital and Training School Depend on Dedicated Staff," Sunday Paper, Charleston, 30 December 1979, Section B, pp. 1–2, Banks Collection, S.C.

20. Anna Banks, "Work of a Small Hospital"; this talk was published in the *Hospital Herald* 1 (May 1899): 8. In her talk Banks claimed that care of the sick in the South had always been the work of black "murmers" or "grannies." See also letters from Banks to Sherman, 14 May 1904, 7 December 1905, 25 April 1910, 13 April 1912, and 18 April 1914; to Miss Cleveland, 22 January 1900; to Dr. Fussell, 10 January 1900; all in Banks Collection, Va.

21. Jane Edna Hunter, *A Nickel and a Prayer* (Cleveland: Elli Kani Publishing, 1940), 52–65. See also Banks, "Work of a Small Hospital."

22. For an excellent discussion of this complex practice of hiring out, see Hine, *Black Women in White*, 47–62.

23. Ibid.

24. Letters from Anna Banks to Miss Davis, 14 May 1900; to Dr. Fussell, 10 January 1900; to Dr. H. B. Fussell, 30 April 1917; to Miss Sherman, 18 April 1914; all in Banks Collection, Va.

25. The *Nurses Quarterly* was published by the Graduate Nurses Association, formed in 1907 to work toward licensure in South Carolina. Initially, all members were graduates of Roper Hospital Training School, but when nurses from other schools began to join, the Roper Hospital Alumnae Association was founded, in about 1910. In 1922, the Graduate Nurses Association formed an official registry to help counteract "favoritism by the doctors" in selecting nurses for their cases. See Chamberlin, *School of Nursing*, 17, 94; "The Visiting Nurses Work in the South" (newspaper report of the Annual Meeting), 1908; Minute Book, 1906–1911, 27–29; Report of the Board, 19 June 1907; Nurse Committee Report, 19 June 1907; all in LBS Collection.

26. The letter, from J. C. Sosnowski, M.D., Secretary of the Medical Society of South Carolina, was signed by twelve physicians. The physicians even went as far as passing a resolution to support the work of the LBS and requesting the society to continue its most "estimable charity." See Minutes of the Medical Society of South Carolina, 1 July 1907, Waring Historical Library, Medical University of South Carolina, Charleston. Although Banks was the subject of their correspondence, she remained unnamed.

27. See, for example, Record Books, 1911 through 1914; Minutes of the Board, 17 May 1912 and 1 January 1923 to 1 June 1923; all in LBS Collection. At the 1916 board meeting, three board members (Poppenheim, Mure, and Cohen) insisted on hiring a white nurse but were outvoted. Poppenheim would become superintendent when Ravenel died in 1933. In 1924, there were 365 black public health nurses in the United States; 158 worked in New York City.

28. The section heading is from Anna DeCosta Banks's "Report of the Visiting Nurse (Colored) to the Ladies Benevolent Society for the Hopkins Fund" (LBS Collection), in which she took the liberty of adding the introductory message, "Open the door of your

heart, my friend, heedless of class or creed when you hear the cry of a brother's voice or the sob of a child in need."

29. Banks's ability to supply "extra nurses of her own color easily" was mentioned in Board Minutes, 19 June 1907; the reduced rates in Board Minutes, 12 April 1909; and the free emergency care in Board Minutes, 1 February 1909; all in LBS Collection. See also *Hospital Herald* 1 (October 1899): 12; Hine, *Black Women in White*, 50–58.

30. Various documents give contradictory information about when the LBS actually began to care for black patients, but by 1835 the society was the recipient of funds from a trust established by John M. Hopkins for the relief of sick and infirm persons of color residing in the city. See the Hopkins will, LBS Collection.

31. Banks's comments are in response to the request that she work for the LBS in 1908. Anna Banks to Catherine Ravenel, Minutes of the Society, Record Book, 1906–1911, LBS Collection. See also Annual Reports, 23 January 1909, 1910, and 1911; Report of the Nurse Committee, 1910; all in LBS Collection; and "Death," *Southern Workman* 60 (February 1931): 94; Report of the Visiting Nurse (Colored), LBS Collection. Ravenel's other editorial comments on Banks's report included adding "colored sick visited" and "colored" to the visiting nurse's title.

32. Banks to Ravenel, Minutes, 1906–1911; "Death."

33. By 1928, the MLI had 2.5 million black policyholders. See Louis Dublin's "The Health of the Negro," *Opportunity* 6 (July 1928): 198; "The Effect of Life Conservation on the Mortality of the Metropolitan Life Insurance Company: A Summary of the Experience, Industrial Department, 1914, Superintendents, Medical Examiners, and Visiting Nurses" (New York: MLI, 1916); "The Effect of Health Education on Negro Mortality," *Opportunity* 2 (August 1924): 232–34; and *After Eighty Years: The Impact of Life Insurance on Public Health* (Gainesville: University of Florida Press, 1966), 148–53. See also Lee Frankel, "A Study of Mortality Statistics of Southern Communities" (paper presented at the Southern Sociological Congress, April 1916). Although 12.3 percent of policyholders were black, they accounted for 16.9 percent of the claims in 1914. By 1923, there were 1.8 million black policyholders; by 1928, 2.5 million. For a more extensive discussion of the MLI nursing service, see Diane Hamilton's "The Cost of Caring: The Metropolitan Life Insurance Company's Visiting Nurse Service, 1909–1953," *Bulletin of the History of Medicine* 63 (fall 1989): 414–34; and "Faith and Finance," *Image, Journal of Nursing Scholarship* 20 (fall 1988): 124–27.

34. A Miss Foute spoke about Jacksonville and a Mrs. Warner explained North Carolinian views at the First Annual Meeting of the NOPHN, June 1913, NOPHN Collection. The Nashville and Birmingham comments are from Ethel Johns, "A Study of the Present Status of Negro Women in Nursing, 1925," Exhibits O-16 and 30, Rockefeller Foundation Archives, box 122, record group 1.1, series 200C. The Richmond plan was presented by Nannie Minor, chief nurse of the Instructional VNA of Richmond (IVNA), at the First Annual Meeting of the NOPHN. For a more extensive discussion of the Richmond story, see Sandra Taylor, "The Role of Black Women in Nursing: A Specific Discussion of the IVNA, Richmond, VA" (paper presented at the American Association for the History of Nursing, 1988).

Jim Crow laws varied over time and place. See Gamble, "Negro Hospital Renaissance";

C. Vann Woodward, *The Strange Career of Jim Crow* (New York: Oxford University Press, 1974), 97–102; Rabinowitz, *Race Relations*. For the more recent debate on Jim Crow laws, see Walter Fraser, Jr., and Winifred B. Moore, eds., *From Old South to the New: Essays on the Transitional South* (Westport, Conn.: Greenwood, 1981); Pauli Murray, *States' Laws on Race and Color; and Appendices* (Cincinnati: Woman's Division of Christian Service, Board of Missions and Church Extension of the Methodist Church, 1955).

35. Minutes of the Board, 9 June 1911, and Minutes from the Annual Meeting, 17 January 1912, LBS Collection. Death rates for black and white MLI policyholders in Charleston were significantly higher than MLI rates nationally; at the same time, Charleston's white policyholder death rate was the highest among all white policyholders in the South. See Lee Frankel, "A Study of Mortality Statistics of Southern Communities" (paper presented at the Southern Sociological Congress, April 1916). For blacks in Charleston the death rate was nineteen per thousand, compared with the national MLI rate of seventeen per thousand. The death rate for Charleston's white policyholders was fourteen per thousand compared with 10.6 per thousand nationally.

36. Anna D. Banks to Miss Sherman, 13 April 1912, and Anna D. Banks to Miss Davis, 18 April 1914, Banks Collection, Va.

37. Cecelia Trescott, a graduate of Harlem Hospital, was the second nurse hired to assist Banks. Book of Annual Reports, 1920–1922, Minutes of the Board, 6 May 1921, LBS Collection.

38. Annual Report 1920, LBS Collection. In 1912, the first complete year with the MLI services, the LBS cared for 238 patients (184 white, 54 black). Banks made 1,919 visits: 1,605 to white patients and 314 to black patients. The fifty-seven MLI patients (40 white, 17 black) received a total of 566 visits. The budget was $1,325, of which $1,000 was from donations. In 1922, there were 1,069 patients and 3,679 visits; the annual budget was $3,333. Metropolitan contributed $1,879, and $1,006 came from donations.

39. The Minutes of the Board, 10 June 1921 (LBS Collection), mention this rumor. The Metropolitan staff who made the visit were a Mrs. Minnie Bridges and a Mr. Henry Mann; see Minutes of the Board, 5 January 1923, and Record Book, 1918–1923, LBS Collection. Metropolitan Life wrote to Ravenel in December 1922 "reporting the decline in death rate and requesting that the Benevolent Society publicize this as widely as possible." This is mentioned in Minutes of the Society, 1 January 1923 to 1 May 1923, 226, and Record Book, 1918–1923, LBS Collection.

40. See "Essentials of Group Nursing, Suggestions for Nurses Interviewing Executives of Metropolitan Group Policyholders," n.d., MLI Collection, folder 160606. Standards for calculating visits per month are outlined in the form letter used by the MLI to initiate new services. The company expected one to three cases per thousand policyholders per month, with each receiving five to six visits per month. Therefore, ten thousand policyholders would require an estimated fifty to 150 visits per month. See, for example, form letter no. 2, Nursing Bureau, Part I, MLI Collection. It is mentioned that the MLI had twelve thousand policyholders in Charleston. See Minutes of the Society, 18 January 1922, LBS Collection. Banks first mentions the addition of group policies in her letter to Miss Davis, 18 April 1914, Banks Collection, Va.

41. Minutes of the Board, 3 February 1923, LBS Collection.

42. Report of the LBS, 1 January 1923 to 1 May 1923, and Annual Report, 17 January 1923, LBS Collection. The visiting nurse was described as a bridge between the classes and masses in Fulmer, "History of Visiting Nurse Work," 412.

43. See letter from Miss M. C. Bull to Rosa H. Clarke, 4 August 1937, quoted in Rosa H. Clarke, *History and Development of Public Health Nursing in South Carolina* (South Carolina State Board of Health, February 1942), 13–14, Waring Historical Library. Bull had been secretary of the LBS board since 1912. At the time, the LBS was working with thirty-five white and nine black doctors. See Annual Report, 16 January 1923, LBS Collection.

44. Woodward, *Strange Career of Jim Crow*, 31.

45. See Reports of the Annual Meeting for 1924–1925, LBS Collection.

46. The Ravenels were Huguenots. In *Piazza Tales: A Charleston Memory* (Charleston: Shaftesbury Press, 1952), Rose Pringle Ravenel, sister of Catherine Ravenel, says, "The definition of a Huguenot is one with 'firm principles and compact prejudices.'"

47. Miriam Ershkowitz and Joseph Kikmund II, *Black Politics in Philadelphia* (New York: Basic Books, 1973), 7–14.

48. Roger Lane, *Roots of Violence in Black Philadelphia, 1860–1900* (Boston: Harvard University Press, 1986), 6–44. In 1900, the largest concentration of black residents was in the seventh ward, bounded on east and west by 7th and 25th Streets and north and south by Spruce and South Streets. See also Allen Davis and Mark Haller, *The Peoples of Philadelphia: A History of Ethnic Groups and Lower-Class Life, 1790–1940* (Philadelphia: Temple University Press, 1973), 176–230.

49. Davis and Haller, *Peoples of Philadelphia*.

50. "Pennsylvania Holds an Inter-Racial Conference," *Opportunity* 2 (February 1924): 51–52. See also Robert Gregg, *Sparks from the Anvil of Oppression: Philadelphia's African Methodists and Southern Migrants, 1890–1940* (Philadelphia: Temple University Press, 1993); Joe Trotter, "Pennsylvania's African American History: A Review of the Literature," in Joe Trotter and Eric Smith, eds., *African Americans in Pennsylvania: Shifting Historical Perspectives* (University Park: Pennsylvania State University Press, 1997), 1–39; Frederic Miller, "The Black Migration to Philadelphia: A 1924 Profile," *Pennsylvania Magazine of History and Biography* 108 (July 1984): 315–50.

51. Commonwealth of Pennsylvania, Department of Welfare, *Negro Survey of Pennsylvania* (Harrisburg, Pa.: Department of Welfare, 1927). See also V. P. Franklin, "The Philadelphia Race Riot, 1918," *Pennsylvania Magazine of History and Biography* 99 (July 1975): 336–50; Bernard J. Newman, "The Housing of Negro Immigrants in Pennsylvania," *Opportunity* 2 (February 1924): 46–48. For a survey of housing conducted by the Armstrong Association, with a slightly more positive analysis, see A. L. Manley, "Where Negroes Live in Philadelphia," *Opportunity* 1 (May 1923): 10–15. The survey had been proposed by Dr. Porter at the January conference; see "Pennsylvania Holds Inter-Racial Conference."

52. "Pennsylvania Holds Inter-Racial Conference."

53. Annual Report 1922, 2, and Minutes of the Board of Managers, 12 October 1923, VNSP Collection. For an excellent discussion of the health of southern blacks and the impact of northern migration, see Edward Beardsley, *A History of Neglect: Health Care*

for Blacks and Mill Workers in the Twentieth-Century South (Knoxville: University of Tennessee Press, 1987), 11–41. Smallpox is discussed in Gregg, *Sparks from the Anvil of Oppression,* 31; Henry Minton, "Negro Physicians and Public Health Work in Pennsylvania," *Opportunity* 2 (March 1924): 74; and "Negro Survey of Pennsylvania," 46–47, in which Forrester Washington contends that his survey committee found that the majority of southern migrants sick with smallpox claimed they had never been told they needed to be vaccinated.

54. That Landis had developed tuberculosis while an intern at Philadelphia Hospital probably explains some of his zeal. Death rates from tuberculosis among blacks decreased from 608 to 362 per hundred thousand between 1910 and 1922. See Harvey Dee Brown, "Tuberculosis Work among Negroes in Philadelphia," *American Review of Tuberculosis* 36 (December 1937): 787–98; David McBride, *Integrating the City of Medicine: Blacks in Philadelphia Health Care, 1910–1965* (Philadelphia: Temple University Press, 1989), 33–55; Barbara Bates, *Bargaining for Life: A Social History of Tuberculosis, 1876–1938* (Philadelphia: University of Pennsylvania Press, 1992), 296–99; Frank Craig, *Early Days at Henry Phipps Institute for the Study and Prevention of Tuberculosis* (Philadelphia: University of Pennsylvania Press, 1952); Fannie Eshleman, "The Negro Nurse in a Tuberculosis Program," *PHN* 27 (July 1935): 375–78.

55. "A Better Chance for Life" was the final heading of the section on health in "Negro Survey of Pennsylvania," 49–52. The components of this better life are described as improved economic status, more and better housing, and health education.

56. Johns, "Present Status of Negro Women in Nursing." Johns visited twenty-five black hospitals and nursing schools in both northern and southern cities. The study was never published, but the findings were disheartening. For an excellent analysis of this study, see Darlene Clark Hine, "The Ethel Johns Report: Black Women in the Nursing Profession, 1925," *Journal of Negro History* 67 (fall 1982): 212–28.

57. Hine, "Ethel Johns Report"; McBride, *Integrating the City of Medicine,* 16–30. Virginia Alexander's article, "Negro Hospitalization" (*Opportunity* 15 [August 1937]: 231–32, 248), was a summary of her proposal to the Milbank Memorial Fund to establish an interracial public health demonstration: "The Social, Economic, and Health Problems of North Philadelphia Negroes and Their Relationship to a Proposed Interracial Public Health Demonstration Center," 14 October 1935, Virginia Mary Alexander Collection, University of Pennsylvania Archives.

58. Vanessa Northington Gamble, *Making a Place for Ourselves: The Black Hospital Movement, 1920–1945* (New York: Oxford University Press, 1995), xv–xvi, 3–34. See also Hine, *Black Women in White,* 27–41; Bates, *Bargaining for Life,* 288–310; McBride, *Integrating the City of Medicine,* 12–55; Gregg, *Sparks from the Anvil of Oppression,* 30–31.

59. "Negro Survey of Pennsylvania," 44–52; Michael Davis, "Problems of Health Services for Negroes," *Journal of Negro Education* 6 (July 1937): 438–49.

60. An article written by Mabel Jacques, a white tuberculosis nurse employed by the VNSP, provides the only public comment on black patients. Jacques describes her black patients as "absolutely frustrating" and the "Philadelphia Negro [as] . . . insolent and overbearing, with a smattering of education to mingle with the superstitions and prejudices of his race, and constantly on the defensive against any suggestion regarding his

mode of living that may benefit him." Mabel Jacques, "The Visiting Nurse in Tuberculosis: Her Importance as an Educational Agent," *Journal of Outdoor Life* 6 (May 1909): 136–37. See also Bates, *Bargaining for Life*, 295.

61. Annual Report 1922, 2; Minutes of the Board of Managers, 12 October 1923, 25 February 1927, and 4 March 1927; all in VNSP Collection. By 1927, only 33 percent of patients were listed as born outside the United States. "Colored" patients were now 23 percent of the caseload. See Annual Report 1927, 12, VNSP Collection.

62. Minutes of the Board of Managers, 25 February 1927 and 4 March 1927, VNSP Collection; "Negro Survey of Pennsylvania," 82. The visiting nurses thought that the black caseload should be representative of the black community. Therefore, in 1920, when 7 percent of the Philadelphia population was black, the VNSP's black caseload was within acceptable limits. This formula was suggested by MLI statistician Louis Dublin, who had given considerable thought to the health problems of the black community. See Louis Dublin, "Records of Public Health Nursing and Their Service in Case Work, Administration, and Research," *PHN* 13 (1921): 285–92.

63. Johns, "Present Status of Negro Women in Nursing," Exhibit E, 1–16, 29–35.

64. Ibid., 29–35.

65. Ibid., 1–16.

66. "For the Health of a Race" is the title of a fundraising pamphlet produced by Mercy Hospital in an effort to build a new nurses' residence. Mercy Hospital Collection, Center for the Study of the History of Nursing, University of Pennsylvania.

67. Johns, "Present Status of Negro Women."

68. Brown, "Tuberculosis Work"; "Pennsylvania Inter-Racial Conference"; Minton, "Negro Physicians," 73–74; Craig, *Early Days at Phipps;* Sadie T. Mossell, *A Study of the Negro Tuberculosis Problem in Philadelphia* (Philadelphia: Henry Phipps Institute, 1923); Bates, *Bargaining for Life,* 291–302; Gamble, *Making a Place;* McBride, *Integrating the City of Medicine;* Jessie M. Robbins, "Class Struggle in the Tubercular World: Nurses, Patients, and Physicians, 1903–1915," *Bulletin of the History of Medicine* 71 (fall 1997): 412–34.

69. Stillman House was supported by Edward Harkness and his sister Charlotte Stillman as an expression of gratitude to a black nanny who had cared for them as children. For an excellent review of Elizabeth Tyler's life, see Marie O. Pitts Mosley, "Satisfied to Carry the Bag: Three Black Community Health Nurses' Contributions to Health Care Reform, 1900–1937," *Nursing History Review* 4 (1996): 65–82; and Adah B. Thoms, *Pathfinders: A History of the Progress of Colored Graduate Nurses* (New York: Kay Printing House, 1929), 40–44.

70. Brown, "Tuberculosis Work."

71. Ibid.; McBride, *Integrating the City of Medicine,* 46–49; Bates, *Bargaining for Life,* 296–99.

72. Bates, *Bargaining for Life,* 296–99; Fannie Eshleman and Marian Dannenberg, "Tuberculosis Training for Colored Student Nurses," *PHN* 15 (June 1923): 301–3; Eshleman, "Negro Nurse," 375–78.

73. Minton, "Negro Physicians," 73–74.

74. McBride, *Integrating the City of Medicine;* Gamble, *Making a Place.*

75. These remarks about black nurses included statements such as: not practical, not

necessary, found them unsatisfactory, not enough Negro patients, patients object. Local Association of Colored Graduate Nurses of Philadelphia and Vicinity, "Summary of Survey of Negro Nurses in Philadelphia," March–June, 1946, Mercy-Douglass Hospital Collection.

76. The experiment in black nurses caring for white patients occurred at Henry Street Settlement in New York; see Johns, "Present Status of Negro Women in Nursing," A-18.

For the Philadelphia situation, see Local Association of Colored Graduate Nurses, "Summary of Survey of Negro Nurses." In Chicago, the VNA hired its first black nurse in 1905. Fulmer asked the white nurses to vote on hiring a black nurse and was disappointed when they vetoed the idea. Over the next few days, however, each nurse came to Fulmer and retracted her negative vote. See Annual Report 1905, 40, VNAC Collection. Most, if not all, of the black nurses were graduates of Provident Hospital. By 1928, the Chicago VNA had nine black nurses on a staff of 120 nurses. See Annual Report 1928, 33, VNAC Collection.

In New York, by 1925 Lillian Wald had a staff of 150 nurses at Henry Street Settlement; twenty-five of these nurses were black. No black nurses were supervisors and none were sent to white homes. See VNSNY, 26 April 1929, Wald Collection, NYPL; Hine, *Black Women in White*, 101

For the situation in the country as a whole, see Louise Tattershall, "Census of Public Health Nursing in the United States, 1924," *PHN* 18 (May 1926): 245–312. The 1931 census found that 62 percent of black public health nurses worked for municipal boards of health or education and 21 percent for VNAs. Thirteen black nurses were employed by insurance companies, mostly in the South (Virginia, South Carolina, and Georgia). See Louise Tattershall, "Census of Public Health Nursing in the United States, 1931," NOPHN Collection; Gamble, *Making a Place*, 105–15; Stanley Rayfield, "A Study of Negro Public Health Nursing," *PHN* 12 (October 1930): 525–36.

77. Cleveland's VNA hired its first black nurse in 1906, ten years before the city's Health Department hired a black nurse. The contract created a "tentative arrangement, which may dissolve at any time if it is found impossible to conduct the work on these lines." A portion of the nurse's salary was obtained from the black business community. See Irene Bower, *Public Health Nursing in Cleveland, 1895–1928* (Cleveland: Cleveland Visiting Nurse Association, 1929), 34; Annual Report, 12 November 1906, and Minutes of the Board, 6 October 1908, 1, VNA of Cleveland.

Richmond's VNA hired its first black nurse in 1910, to work under the supervision of black physicians. The board sought funds for her salary from the black community, threatening to fire her when contributions lagged. By 1923, seventeen white and seven black nurses were on the staff. See Report of the Negro Welfare Committee, *The Negro in Richmond* (Richmond: Richmond Council of Social Agencies, 1929), 64; "The Nurses' Settlement of Richmond, Va.," n.d., 8, and Annual Report 1930s, VNAR Collection. Norfolk followed a similar plan. See "Report of the Convention," Graduate Nurses Association of Virginia, 1911, 45.

78. For a discussion of the challenges of private duty in Charleston and Cleveland, see Hunter, *Nickel and a Prayer*; Massey Riddle, "Source of Supply of Negro Personnel: Section C: Nurses," *Journal of Negro Education* 6 (July 1937): 483–92; Abbie Roberts, "Nurs-

ing and Opportunities for the Colored Nurse," *Proceedings of the National Conference of Social Work* (Chicago: University of Chicago Press, 1928), 183–85; Davis, "Problems of Health Services for Negroes," 443; Hospital Library and Service Bureau, *Report on the Informal Study of the Educational Facilities for Colored Nurses and Their Use in Hospitals, Visiting Nursing, and Public Health Nursing* (Chicago: Hospital Library and Service Bureau, 1924–25), reprinted in Darlene Clark Hine, *Black Women in the Nursing Profession: A Documentary History* (New York: Garland, 1985), 45–59; Rayfield, "Study of Negro Public Health Nursing"; Darlene Clark Hine, "They Shall Mount up with Wings as Eagles: Historical Images of Black Nurses, 1890–1950," in Ann Hudson Jones, ed., *Images of Nurses: Perspectives from History, Art, and Literature* (Philadelphia: University of Pennsylvania Press, 1988), 177–96.

79. See Hine, "Ethel Johns Report," 219–20; Nannie Minor, "Status of Colored Public Health Nurses in Virginia," *PHN* 16 (May 1924): 243; Frank O. Nichols, "Opportunities and Problems of Public Health Nursing Negroes," *PHN* 16 (March 1924): 122; Rayfield, "Study of Negro Public Health Nursing."

80. *Statistical Bulletin, MLI* 8 (August 1927): 8–9. This story about race adjustment is found in Dublin, "Health of the Negro," 200–216. White nurses' descriptions of their patients certainly conveyed a message of disdain and inferiority. See Sarah Meyers, "The Negro Problem as It Appears to a Public Health Nurse," *AJN* 19 (January 1919): 278–81; Ann Doyle, "Rural Nursing among Negroes," *PHN* 12 (December 1920): 981–85; Gamble, "*Making a Place,*" 7; Vanessa Northington Gamble, *Germs Have No Color Line: Blacks and American Medicine, 1900–1940* (New York: Garland, 1989). The "weakest link" comment is from a graduation address at Harlem School of Nursing by Evelyn Pitter, "The Colored Nurse in Public Health," *PHN* 26 (September 1926): 719. The expression "a race's problem" is from the fundraising brochure, "For the Health of a Race." See also Johns, "Present Status of Negro Women in Nursing," 16, 30.

81. Elizabeth Jones, "The Negro Woman in the Nursing Profession," *Messenger* 5 (July 1923): 764–65; Hine, "They Shall Mount Up," 177.

82. Jessie Marriner, "Public Health Nurses for the Negro Race in Alabama," *PHN* 15 (June 1923): 306.

83. Johns, "Present Status of Negro Women in Nursing," 51. For a discussion of educational strategy, see Rayfield, "Study of Negro Public Health Nursing," 526–28. The ten programs that admitted black nurses are listed in Pitter, "Colored Nurse."

84. Rayfield, "Study of Negro Public Health Nursing"; Hine, *Black Women in White,* 47–62.

85. Rayfield, "Study of Negro Public Health Nursing"; Henrietta Landau, "Registered Negro Nurses in the USA," *AJN* 43 (August 1943): 730–33. White philanthropies of the Julius Rosenwald Fund did take on black health care as an important agenda. The fund provided financial incentives to health departments to hire black personnel and supported demonstration projects to improve black health care. Under the auspices of the Negro Health Division, numerous employment opportunities opened up for black nurses, especially in the South. See Gamble, *Making a Place,* 109–10.

86. The Nurse Volo story is recorded in "Summary of Visits to Schools of Nursing," Interviews: Mary Beard, 11 December 1926, Mary Beard Collection, Department of Manuscripts and University Archives, Cornell University Libraries.

87. Mary Beard, "Nurse Emma," *Survey* 62 (April 1929): 113–14. This story was also included in Beard's book *The Public Health Nurse* (New York: Harper and Brothers, 1929), written for use by administrators and instructors. See also Jones, "Negro Woman."

88. Gloria R. Smith, "From Invisibility to Blackness: The Story of the National Black Nurses' Association," *Nursing Outlook* 23 (April 1975): 226; Louise P. Nelson, "My Part," *AJN* 28 (April 1928): 355. Louise Nelson was an administrator at Freedmen's Hospital in Washington, D.C.

89. Mosley, "Satisfied to Carry the Bag"; Hine, "They Shall Mount Up." See also Mabel Staupers, *No Time for Prejudice: A Story of the Integration of Negroes in Nursing in the United States* (New York: Macmillan, 1961), 7; M. Elizabeth Carnegie, *The Path We Tread: Blacks in Nursing Worldwide, 1854–1994* (New York: National League for Nursing, 1995), 154–55; Adah B. Thomas, *Pathfinders: A History of the Progress of Colored Graduate Nurses* (New York: Kay Printing House, 1929).

90. Sleet's story is in Thomas, *Pathfinders*.

91. J. C. Sleet, "A Successful Experiment," *AJN* 1 (July 1901): 729–31.

92. Ibid.; J. S. Scales, "Tuberculosis among Negroes: A Report to the Committee on the Prevention of Tuberculosis (1904–05)," included in the Third Annual Report of the Committee on the Prevention of Tuberculosis of the Charity Organization Society of the City of New York; Mosley, "Satisfied to Carry the Bag."

93. Scales, "Tuberculosis among Negroes."

94. See, for example, Dublin, *After Eighty Years,* 148–53; Louis Dublin, "Recent Improvement in the Negro's Mortality," *Opportunity* 1 (April 1923): 5–8; Eugene Kinckle Jones, "The Negro's Struggle for Health," *Opportunity* 1 (June 1923): 4–8; C. C. Spaulding, "Improvements in Negro Health as Shown by Insurance Records," *Opportunity* 1 (December 1923): 364–66.

95. Spaulding, "Improvements in Negro Health."

96. M. O. Bousfield, "Reaching the Negro Community," *AJPH* 24 (1934): 209–33. See also Rayfield, "Study of Negro Public Health Nursing."

97. See, for example, Kevin Schulman, Jesse Berlin, William Harless, et al., "The Effect of Race and Sex on Physicians' Recommendations for Cardiac Catheterization," *New England Journal of Medicine* 340 (February 1999): 618–26.

Chapter 5: Lillian Wald and the Invention of Public Health Nursing

1. VNSNY, *Healing at Home: Visiting Nurse Service of New York, 1893–1993* (New York: VNSNY, 1993); Karen Buhler-Wilkerson, "Bringing Care to the People: Lillian Wald's Legacy to Public Health Nursing," *AJPH* 83 (December 1993): 1778–86; Josephine Goldmark, *Nursing and Nursing Education in the United States* (New York: Macmillan, 1923): 42–43. Wald's comment on the role of the nurse is from VNSNY, *Healing at Home.*

2. Robert Bremer, "Lillian Wald," in Edward James, Janet James, and Paul Boyer, eds., *Notable American Women: A Biographical Dictionary* (Cambridge: Belknap Press of Harvard University Press, 1971), III: 526–29; Marguerite Wales, *The Public Health Nurse in Action* (New York: Macmillan, 1941), xi; Lillian Wald, "We Called Our Enterprise Public Health Nursing," in Wales, *Public Health Nurse,* xi, xiii; Lillian Wald, *The House on Henry Street* (New York: Henry Holt, 1915). Wald discussed the nurse's organic relationship with

the community in "Visiting Nursing" (unpublished paper), 11 December 1918, Wald Collection, NYPL, and "What I Would Do with a Million Dollars," 26 December 1918, Wald Collection, roll 25, NYPL.

3. Several historians have written about Wald's life. See, for example, Doris Groshen Daniels, *Always a Sister: The Feminism of Lillian D. Wald* (New York: Feminist Press, 1989); Robert Duffus, *Lillian Wald: Neighbor and Crusader* (New York: Macmillan, 1938); Beatrice Siegel, *Lillian Wald of Henry Street* (New York: Macmillan, 1983).

4. Wald's comment is from her application to training school; see Lillian Wald to George Ludlum, 27 May 1889, Wald Collection, roll 1, NYPL. See also Duffus, *Lillian Wald,* 23–26.

5. Wald, *House on Henry Street,* 1–5. See also Duffus, *Lillian Wald,* 31–40.

6. Lavinia L. Dock, "Whence the Term Public Health Nursing?" *PHN* 29 (1937); Duffus, *Lillian Wald,* 34–35; Wald, *House on Henry Street.*

7. Wald, *House on Henry Street,* 3; Allen Resnick, "Lillian Wald: The Years at Henry Street" (Ph.D. diss., University of Wisconsin, 1973), 83, n. 11; Daniels, *Always a Sister,* 70; Duffus, *Lillian Wald,* 35–40. These accounts do not completely agree on the events of this period. See also Wald, *House on Henry Street,* 24.

8. Lillian Wald, Report "B," 2 July 1893, VNSNY Collection. This was one of Wald's early reports to Schiff. She apparently also kept careful account books, recording all expenses.

9. Henry Street's ledger during the first decade of the twentieth century lists only a small group of consistent individual and institutional contributors. In contrast, most VNAs' lists of benefactors cover many pages. Wald's early benefactors included Arnstein, Bar, Bliss, Blumenthal, Bowdoin, Brown, Cohn, Harbener, Harkness, Heinsheimer, Herrmanne, Lewisohn, Loeb, Morganthau, Mortan, Ripley, Schiff, Whitney, Wile, and "a friend." See "Henry Street Ledger, 1903–1909," VNSNY Collection. For an analysis of the funding of Hull House (settlement house founded by Jane Addams in Chicago), see Kathryn Kish Sklar, "Who Funded Hull House?" in Kathleen McCarthy, ed., *Lady Bountiful Revisited: Women, Philanthropy, and Power* (New Brunswick, N.J.: Rutgers University Press, 1990), 94–115. See also Clare Coss, *Lillian D. Wald: Progressive Activist* (New York: Feminist Press, 1989), 1–15; Siegel, *Lillian Wald,* 89–105.

10. Duffus, *Lillian Wald,* 34; Dock, "Whence the Term?"

11. For a more extensive discussion, see Karen Buhler-Wilkerson, *False Dawn: The Rise and Decline of Public Health Nursing, 1900–1930* (New York: Garland Press, 1990), 14–19; Dock, "Whence the Term?"; Lavinia Dock, "The History of Public Health Nursing," *PHN* 14 (1922): 525; Wald, *House on Henry Street;* Duffus, *Lillian Wald,* 40.

We know that Wald attended the Chicago Fair because her questions during the discussions are recorded. See John Billings and Henry M. Hurd, eds., *Hospitals, Dispensaries, and Nursing: Papers and Discussions in the International Congress of Charities, Corrections, and Philanthropy, Section III, Chicago, June 12–17, 1893* (Baltimore: Johns Hopkins University Press, 1894), 524, 525. On Craven's and Nightingale's influence, see Ellen C. Lagemann, *A Generation of Women: Education in the Lives of Progressive Reformers* (Cambridge: Harvard University Press, 1979), 68, 71–78; and Resnick, *Lillian Wald,* 62–68. On Dock's correction and editing of Wald's writing, see, for example, Wald to Dock, 18 February 1915, Wald Collection, roll 25, NYPL.

12. Wald, *House on Henry Street*, 9, 26–27; Wald, "Visiting Nursing"; Carroll Smith-Rosenberg, *Disorderly Conduct: Visions of Gender in Victorian America* (New York: Oxford University Press, 1985), 245–96. For a discussion of gender and the development of public health nursing, see Barbara Melosh, *The Physician's Hand: Work, Culture, and Conflict in American Nursing* (Philadelphia: Temple University Press, 1982); Martha Vicinus, *Independent Women: Work and Community for Single Women, 1850–1920* (Chicago: University of Chicago Press, 1985); Kathryn Kish Sklar, *Florence Kelly and the Nations: The Rise of Women's Culture, 1830–1900* (New Haven: Yale University Press, 1995).

13. Wald, *House on Henry Street*, 27–28; Wald, "Visiting Nursing."

14. Wald, *House on Henry Street*; Lavinia Dock, "As the Nurse Sees It," *Charities and Commons* 16 (April 1906): 10–12; Lillian Wald, "Nurses' Settlement," *AJN* 1 (June 1901): 682–84.

15. Wald, "We Called Our Enterprise," xi, xiii; Lillian Wald, *Windows on Henry Street* (New York: Henry Holt, 1915). Wald claimed to have originated the term *public health nurse* in 1893. See Lillian Wald, "Development of Public Health Nursing in the United States," *Trained Nurse* 80 (June 1928): 689–92. Wald on several occasions wrote to correct the historical record when she was not properly credited. See, for example, Lillian Wald to Editor, *Red Cross Courier*, 13 May 1922; and Wald to Dockie Darling (Lavinia Dock), n.d.—both in Wald Collection, roll 12, NYPL; Wald to Katharine Tucker, 27 October 1934, Wald Collection, NYPL. See also Dock, "As the Nurse Sees It," 10–12; Wald, "Nurses' Settlement," 682–84.

16. Dock claimed that the nurses issued no public or formal reports and that no appeals were made for money. See Lavinia Dock, "The Nurses' Settlement in New York," n.d. [appears to have been written around 1899], Wald Collection, Columbia University.

Initially, Wald and Brewster kept "Daily Records," but by October 1893 the ravages of the depression had created such great demand on their time that Wald wrote in a letter to Schiff and Loeb that "very little record of the work beyond addresses has been possible." For the rest of the nurses' time on Jefferson Street, the only remaining records are Wald's monthly letters to her benefactors and a report of expenses, dated January 1895. The remaining letters are in the safe of the VNSNY. Copies of several letters are in the Wald Collection, NYPL: "Daily Record": "A," "B" 2 July 1893, "C" 2–7 July 1893; 24 July 1893, and 25 July 1893; Wald to Jacob Schiff and Mrs. Solomon Loeb, 14 July 1893, 29 July 1893, 2 October 1893, 3 November 1893, 2 February 1894, 3 March 1894, 8 December 1894, 11 February 1895, and 10 April 1895.

17. See all letters cited in n. 16.

18. See all letters cited in n. 16; Wald to Schiff and Loeb, 28 November 1893, 4 January 1894, and 4 March 1894, Wald Collection, NYPL.

19. See all letters cited in n. 16; Bremer, "Lillian Wald," 527. Wald's writing to Schiff that "the best aid we obtain from friends is when they seek and find employment for our neighbors" had apparently contributed to her access to so many jobs. See also Wald to Schiff and Loeb, 1 October 1894; Wald, *House on Henry Street*, 14. Wald and Brewster identified Lowell and the landlady of their Jefferson Street house, Mrs. McRae, as their earliest influences and guides of greatest distinction. Wald described Lowell's genius in Wald to W. R. Stewart, 10 October 1910, Wald Collection, roll 1, NYPL.

20. Wald, "Nurses' Settlement"; Wald, "Development of Public Health Nursing." An-

nie Goodrich, who became director of Henry Street when Wald was no longer able to oversee daily operations, referred to Wald as the Madonna of the Slums; see Daniels, *Always a Sister*, 24.

21. Wald, *House on Henry Street*, 81–88. See also "Daily Record," 2–7 July 1893; and Wald to Schiff and Loeb, 4 January 1894, 1 October 1894, 10 April 1895, and 7 December 1894, Wald Collection, NYPL. Goldman's comment is from Daniels, *Always a Sister*, 85, 143, and is also mentioned in Duffus, *Lillian Wald*, 25, and in Wald to Dock, 4 November 1931, Wald Collection, NYPL.

22. Wald to Schiff and Loeb, 4 January 1894 and 11 February 1895, Wald Collection, NYPL. In January 1895, Wald claimed she and Brewster cared for 125 patients requiring nursing help and an uncounted number of people needing assistance. In addition to salaries, the nurses' expenses for that month were $99.86.

23. Lillian Wald, "Nurses in 'Settlement' Work," *Proceedings of the National Conference of Charities and Corrections*, 24–30 May 1895. Schiff purchased the Henry Street house in January 1895 and increased Wald's stipend. For a more detailed discussion of the "old girl network" that developed at the settlement, see Daniels, *Always a Sister*, 62–74.

24. Daniels, *Always a Sister*. For excellent analyses of this community of women, see Coss, *Lillian Wald*, 5–7; and Resnick, "Lillian Wald," 111–12. See also Wald, *Windows on Henry Street*, 6–10.

25. Dock, "The Nurses' Settlement"; Resnick, "Lillian Wald," 107–76.

26. Wald, *Windows on Henry Street*, 49–50; Daniels, *Always a Sister*, 50. Wald was an early member of the small group that organized the NAACP and a signer of "The Call" to action. See Minnie Finch, *The NAACP: Its Fight for Justice* (Metuchen, N.J.: Scarecrow Press, 1981), 8–12, 252–53; Carolyn Wedin, *Inheritors of the Spirit: Mary White Ovington and the Founding of the NAACP* (New York: John Wiley, 1998), 108.

27. Wald, "Nurses' Settlement"; Daniels, *Always a Sister*, 39–68; *Report of the Henry Street Settlement, 1893–1918* (New York: Henry Street Settlement, 1918); Resnick, "Lillian Wald," 166.

28. The ledger entries for 1 September and 1 October 1909 list a payment from the Hunson Guild for $70.

29. See, for example, Lavinia Dock's "An Experiment in Contagious Nursing," *AJN* 12 (September 1903): 927–32; and "The School Nurse Experiment in New York," *AJN* 3 (November 1902): 109–10. See also Jane Hitchcock, "Five Hundred Cases of Pneumonia," *AJN* 3 (December 1902): 169; Resnick, "Lillian Wald"; Daniels, *Always a Sister;* Siegel, *Lillian Wald;* Bremer, "Lillian Wald," 526–29; Duffus, *Lillian Wald*.

30. Minutes of the Nurse Committee, Henry Street Settlement, 29 September 1911, 15 October 1915, and 14 November 1916; Dock to Wald, 1 February 1904 and 30 June 1904; "Henry Street Settlement: Department of Nursing"; "Visiting Nurse Service Administered by Henry Street Settlement: Bulletin of Instruction, 1920–1921"; all in Wald Collection, Columbia University, Rare Book and Manuscript Library. For statistics, see "Report of the Department of Nursing, 1909–1910" and "Nurse Accounts, 1902–1919," Wald Collection, Columbia University. For a chronology of important events, see "Report of the Visiting Nurse Service Administered by the Henry Street Settlement, 1923," VNSNY Collection; "New Work of the Nurses' Settlement," *AJN* 1 (December 1900): 343–44; Wald, "Nurses' Settlement."

31. See all sources cited in n. 30. Standing orders provided nurses with coverage until a physician could be secured. See, for example, NOPHN, *Manual for Public Health Nursing* (New York: Macmillan, 1927), 25–26; "Standing Orders," *AJN* 13 (March 1913): 451–53; "Standardization of Standing Orders," *Trained Nurse* 55 (December 1915): 357–58.

By 1916, a settlement study found that in some neighborhoods the necessity for first aid rooms was passing. Given all the restrictions and the growing availability of acceptable neighborhood dispensaries, the nurses decided to decrease the service slowly, with the possibility of closing it altogether. By the next year, the number of first aid cases dropped to 12,428.

32. "Report of Department of Nursing, 1909–1910"; "Nurse Accounts, 1902–1919"; "Report of Visiting Nurse Service, 1923"; Henry Street Settlement, "Report of the Nurses' Department of the Henry Street Settlement, 1911," Columbia University.

33. "Report of Department of Nursing, 1909–1910"; "Nurse Accounts, 1902–1919"; "Report of Nurses' Department of Henry Street Settlement, 1911"; VNSNY, *Healing at Home*.

34. For excellent reviews of this growth, see Resnick, "Lillian Wald," 161–89; Henry Street Settlement, *Report, 1893–1918*. See also "Report of Nurses' Department of the Henry Street Settlement for the Years 1909 and 1910" and "Henry Street Settlement: Bulletin of Instruction, 1920–1921," Wald Collection, Columbia University. For further discussion, see Yssabella Waters, *Visiting Nursing in the United States* (New York: Charities Publications, 1909), 29–37; Knight Hurst to Lillian Wald, 3 July 1902, Wald Collection, box 42, "Board Misc.," Columbia University.

35. Resnick, "Lillian Wald," 161–89; Henry Street Settlement, *Report, 1893–1918*.

36. Wald, *House on Henry Street,* 161–63; "Report of the Visiting Nurse Service Administered by the Henry Street Settlement, 1923," "Report of Henry Street Settlement, Visiting Nurse Service Included, 1926," and "Combined Report of the Visiting Nurse Service and Social Activities of Henry Street Settlement, 1927," all in Wald Collection, Columbia University; Marie O. Pitts Mosley, "Satisfied to Carry the Bag: Three Black Community Health Nurses' Contributions to Health Care Reform, 1900–1937," *Nursing History Review* 4 (1996): 65–82.

37. Mosley, "Satisfied to Carry the Bag."

38. Letters from Wald to Josephine Goldmark, 12 May 1936; to Helen Hall, 19 May 1936; to Rita Morganthau, 25 May 1936; to Marguerite Walls, 31 May 1936; to Ellen Buell, 7 July 1936; to Joint Vocational Service, Inc., 29 June 1936; and to Rita Morganthau, 14 September 1936; Minutes, Special Meeting of the Board, 24 May 1944; all in Wald Collection, box 9, roll 7, NYPL.

39. Karen Buhler-Wilkerson, "Henry Street Settlement" and "Visiting Nurse Service of New York," in Kenneth Jackson, ed., *Encyclopedia of New York* (New Haven: Yale University Press, 1995), 540, 1229.

40. Wald is quoted in VNSNY, *Healing at Home*. See also Wald, *Windows on Henry Street,* 10.

Chapter 6: The Business of Private Nursing

1. For excellent histories of private-duty nursing, see Jean Whelan, "Too Many, Too Few: The Supply, Demand, and Distribution of Private-Duty Nurses, 1910–1965" (Ph.D.

diss., University of Pennsylvania, 2000); Susan Reverby, "Neither for the Drawing Room Nor for the Kitchen: Private-Duty Nursing in Boston, 1873–1910," in Judith Leavitt, ed., *Women and Health in America* (Madison: University of Wisconsin Press, 1984), 454–66; Susan Reverby, "'Something besides Waiting': The Politics of Private-Duty Nursing Reform in the Depression," in Ellen Lagemann, ed., *Nursing History: New Perspectives, New Possibilities* (New York: Teachers College Press, Columbia University, 1983), 133–56; Susan Reverby, *Ordered to Care: The Dilemma of American Nursing, 1850–1945* (Cambridge: Cambridge University Press, 1987); Barbara Melosh, *The Physician's Hand: Work, Culture, and Conflict in American Nursing* (Philadelphia: Temple University Press, 1982); Katheryn McPherson, *Bedside Matters: The Transformation of Canadian Nursing, 1900–1990* (Toronto: Oxford University Press, 1996); Darlene Clark Hine, *Black Women in White: Racial Conflict and Cooperation in the Nursing Profession, 1890–1950* (Bloomington: Indiana University Press, 1989), 53–61.

The decision of Philadelphia's College of Physicians to rely on trained nurses to staff its registry is discussed in W. W. Keen, Wharton Sinkler, and James Wilson, "To the Fellows of the College of Physicians of Philadelphia," 1890, CPP Collection, CPP 10/0001-01.

2. Miss Crowell, "The Art of Nursing" (paper presented at the Graduate Nurses Association of Virginia Convention, 24 May 1915), 47, VNAR Collection.

3. Miss Holleman, "The Private-Duty Nurse and Her Work" (paper presented at the Graduate Nurses Association of Virginia Convention, 1915), 43–44, VNAR Collection.

4. Jeannette Forest, "One Hundred Don'ts for Nurses," *Trained Nurse* 13 (October 1894): 190–93. See also Charlotte Kettles, "Sequel to 'The Shall-Be Nurse,'" *Canadian Nurse and Hospital Review* 8 (April 1912): 173–77; Sarah Pierpont, "Hints to Trained Nurses," *Trained Nurse* 25 (August 1900): 105–6; A.D., "Patients' Criticism of Nurses," *AJN* 13 (July 1913): 765–67; Elizabeth P. Preston Cocke, "The Obligations of the Registered Nurse"(paper presented at the Graduate Nurses Association of Virginia Convention, June 1908), VNAR Collection (Preston used the expression "grit, gumption and grace," chosen for the heading of this section); Donna Wills, "The Problem of Private Nursing," and Miss M. Seibert, "Private Nursing in Country Districts" (papers presented at the 1916 Graduate Nurses Association of Virginia Convention), and Jennie Thecla Traylor, "Attention to Detail" (paper presented at the 1905 Graduate Nurses Association of Virginia Convention), VNAR Collection; Minnie Morse, "Nursing in Hotels and Boarding Houses," *Trained Nurse* 55 (November 1915): 293; Holleman, "Private-Duty Nurse"; Bertha J. Gardner, "Private-Duty Emergencies," *AJN* 13 (September 1913): 1005–11.

5. See all sources cited in n. 4; A. E. Potts, "Nursing Typhoid Fever in 'Little Italy'— Philadelphia," *Trained Nurse* 33 (November 1904): 301–4; Katherine DeWitt, *Private-Duty Nursing* (Philadelphia: J. B. Lippincott, 1917), 16–17.

6. Cocke, "Obligations of the Registered Nurse." One nurse described the cook as "John Chinaman." See Emily Woodman, "Some Problems of the Nurse on Private Duty," *Nurses' Journal of the Pacific Coast* 6 (July 1910): 305–9; Margaret Ramsey, "Letters from a Private-Duty Nurse," *AJN* 14 (January 1914): 278–81; Grace Holmes, "Working for Our Living," *AJN* 9 (January 1909): 252–56; Mrs. L. D. Palmer, "The Advantages of Training Nurses in Families," *Trained Nurse* 40 (May 1908): 293–97; A Lakeside Graduate, "Prob-

lems in Private Nursing," *AJN* 6 (June 1906): 597–99; Margaret L. Rogers, "Private Nursing," *Trained Nurse* 27 (October 1901): 187–90.

Florence Nightingale, speaking on private nursing, expressed concern that "the danger is that the private nurse may become an irresponsible nomad. She has no home." See Adelaide Nutting, "Work of the Alumnae Association," *Trained Nurse* 13 (September 1894): 113–16; Elsa Sperry, "The Nurse on Private Duty," *AJN* 16 (June 1916): 854–59. It was said that after about ten years of steady work, most private-duty nurses found their "health and enthusiasm failing"; see, for example, Anne A. Hintze, "On Business Principles," *Trained Nurse* 15 (July 1893): 1–3. See also An Observer, "Is the Profession Becoming Overcrowded?" *AJN* 3 (April 1903): 513–14.

7. Janet Geister, "Private-Duty Nursing, Then—And Now," *Trained Nurse* 100 (April 1938): 384.

8. Frederick P. Gay, *Typhoid Fever: Considered as a Problem of Scientific Medicine* (New York: Macmillan, 1918), 13–24. By 1913, the death rate for typhoid patients decreased by 50 percent. Typhoid was considered a "sanitary index" of purity of water and food supply, and this decline apparently reflected improvements in these areas. Better care and decreased virulence of, or increased resistance to, infection were also thought to contribute to this decline in deaths. With the introduction of antimicrobial therapy (chloramphenicol), the case mortality was reduced to 1 percent. See Paul Beeson, Walsh McDermott, and James B. Wyngaarden, *Cecil: Textbook of Medicine* (Philadelphia: W. B. Saunders, 1979), I: 446–49; Geister, "Private-Duty Nursing," 384.

9. William Osler and Thomas McCrae, *The Principles and Practice of Medicine* (New York: D. Appleton, 1920), 38; S. Virginia Levis, "Communicability and Prevention of Typhoid Fever," *Trained Nurse* 44 (April 1910): 286–90. Levis discusses how nurses were taught that typhoid was "highly infectious" but "in degree contagious" and that nurses therefore never considered a patient a "menace"; see her "Nursing in Typhoid," *Trained Nurse* 15 (September 1895): 124–31. See also George P. Paul, "Nursing in Typhoid Fever," *Trained Nurse* 34 (January 1905): 22–25; Beeson et al., *Cecil*, 446–49. A. L. Benedict, in "Nursing in Typhoid Fever" (*Trained Nurse* 34 [May 1905]: 307–12), contended that he would rather pass through an attack of typhoid with no doctor at all but the care of a good nurse than with the best doctor in the world and an incompetent nurse.

10. Stella Goostray, "The Nursing Care in Typhoid Fever," *AJN* 27 (September 1927): 719–22; Wm. G. Daggett, "The Treatment of Typhoid Fever," *Trained Nurse* 5 (September 1890): 101–8; Levis, "Communicability and Prevention," 286–90; Beeson et al., *Cecil*, 446. For a review of the conception of contagion in typhoid, see Gay, *Typhoid Fever*, 6–7, 39–42. The bacillus was not discovered until 1880. See also DeWitt, *Private-Duty Nursing*, 206–26.

11. Anna R. Nelson, "A Typhoid Case in the Country," *Trained Nurse* 44 (April 1910): 291–97; "Extracts from a Nurse's Journal," *Trained Nurse* 5 (December 1890): 256–59. For an amazing list of what a nurse might carry along, see Helen Hay, "The Outfit of the Private-Duty Nurse," *AJN* 1 (January 1901): 264–66. See also Hattie M. Greaves, "A Nurse's Experience during the Typhoid Epidemic in Butler, Pa.," *Trained Nurse* 32 (June 1904): 381–84. There were 976 cases of typhoid in Butler during the epidemic; nurses (207) and physicians arrived from around the country to help. Sixteen nurses contracted the disease and several died. Four physicians got sick, but only one died.

12. "Extracts from a Nurse's Journal." Anna Nelson, in "Typhoid Case," also described the need to sleep fully dressed when she encountered "the dirtiest, most uninviting place on which to rest." See also "A Nurse's Duties to Herself," *Trained Nurse* 3 (July 1889): 1–3.

13. "A Nurse's Duties to Herself."

14. Inez C. Lord, "Typhoid Fever," *AJN* 3 (January 1903): 276–80; Benedict, "Nursing in Typhoid Fever," 307–12. The incidence of hemorrhage was 7 to 20 percent; approximately 3 percent of typhoid patients experienced perforation and 10 percent pneumonia. According to Osler, 25 to 33 percent of typhoid deaths were due to perforation; see Osler and McCrae, *Principles,* 22–23. See also Beeson et al., *Cecil,* 447; "Notes on Fever Nursing," *Trained Nurse* 2 (January 1888): 1–7, and 3 (December 1889): 216–22.

15. Lord, "Typhoid Fever"; Benedict, "Nursing in Typhoid Fever," 307–12; Osler and McCrae, *Principles,* 22–23; Beeson et al., *Cecil,* 447; "Notes on Fever Nursing."

16. Osler and McCrae, *Principles,* 38–43; Daggett, "Treatment of Typhoid Fever"; Susan Long, "Notes of a Typhoid Fever Case," *Trained Nurse* 6 (March 1891): 126–33; Miss F. Pearse, "Notes on Typhoid Nursing," *Trained Nurse* 4 (October 1891): 145–52; Elizabeth Campbell Gorden, "Some Observations on the Nursing of Typhoid Fever," *AJN* 3 (May 1903): 593–600; Levis, "Communicability and Prevention"; Paul, "Nursing in Typhoid Fever"; Margaret Powell, "Notes on Private Nursing of Infectious Cases," *Trained Nurse* 6 (June 1891): 261–64.

17. Long, "Notes of a Typhoid Fever Case"; Pearse, "Notes on Typhoid Nursing"; Edward B. Angell, "The Physiological Basis of Hydrotherapy," *AJN* 3 (May 1903): 600.

18. See all sources cited in n. 16.

19. Osler and McCrae, *Principles,* 38; Daggett, "Treatment of Typhoid Fever," 103.

20. Gay, *Typhoid Fever,* 132–35; Nelson, "Typhoid Case"; Goostray, "Nursing Care in Typhoid Fever"; DeWitt, *Private-Duty Nursing,* 206–16; Benedict, "Nursing in Typhoid Fever"; Levis, "Typhoid Fever"; "Disinfection in Contagious Diseases and Epidemics," *Trained Nurse* 4 (February 1890): 85–86; Cocke, "Obligations of the Registered Nurse." Other disinfectant agents included carbolic acid, bichloride solution, bichloride of mercury, formalin, carbolic acid, calcium hydroxide (milk of lime), copper sulfate, and ferrous sulfate.

21. Lavinia Dock, "The Question of Nurses' Directories," *Trained Nurse* 8 (October 1894): 171–74; Edward H. Brigham, "History of the Directory for Nurses of the College of Physicians," n.d., CPP Collection, CPP 10/0001-01; Geister, "Private-Duty Nursing." The problems of "foreign nurses" (nurses from out of town) are discussed in Grace Holmes, "An Ideal Central Directory," *AJN* 7 (June 1906): 606–8. See also "Letter to the Editor from Graduate," *Trained Nurse* 14 (January 1895): 97.

22. Dock, "Question"; Geister, "Private-Duty Nursing."

23. "To Physicians from Office of the Washington Directory for Nurses," 15 December 1882, CPP Collection.

24. Dr. James Wilson, "Directory for Nurses of the College of Physicians," n.d. Various newspaper clippings in the Directory for Nurses in the CPP Collection help tell the story. See, for example, "War of the Nurses Shifted to Court," n.d.; "A Rival Directory: Movement on Foot to Create Opposition to the College of Physicians," n.d.; and "Quackery in the Nursing Profession," 1900. The use of a "black book" to chronicle careers of

offenders—nurses who received complaints—is mentioned in a letter written by the registrar of Boston's Directory of Nurses to the Philadelphia Directory; see Edwin H. Brigham to Dear Dr., 26 May 1882, CPP Collection. There were also controversies over the use of nurses certified by the Philadelphia Nurse Supply Association. See also L. L. Dock, "Central Directories and Sliding Scales," *AJN* 7 (1907): 11–13; Sperry, "Nurse on Private Duty."

25. Dock, "Question." Charges of profiteering by untrained nurses employed by a commercial directory in Cleveland caused a great deal of public and professional concern. See "Sift Charges That Nurses Profiteer: Councilmen Call on U.S. Agent to Investigate Bureau," *Cleveland Plain Dealer,* 15 December 1918; "Two Courses Open in Nurse Inquiry: Council Undecided on Policy in Dealing with Profiteering," *Cleveland Plain Dealer,* 20 December 1918. The Cleveland Registry also received a visit (in 1910) from the Attorney General's Office because the registry was running an employment bureau without paying the necessary $100 registration fee to the Labor Commission. The registrar was able to avoid arrest by claiming that her organization was "nonprofit." See "History of the Registry," 1: 3, and "Report of Committee on Recommendations Following Investigations of Profiteering by Nurses," 1918, 4: 6, GCNA Collection.

26. Dock, "Question"; Committee for the Study of Nursing Education, *Nursing and Education in the United States* (New York: Macmillan, 1923), 161–83; May Ayes Burgess, *Nurses, Patients, and Pocketbooks: A Study of the Economics of Nursing by the Committee on the Grading of Nursing Schools* (New York: Committee on the Grading of Nursing Schools, 1928), 33–89; The Observer, "Some Impressions of the San Francisco Convention," *Trained Nurse* 55 (October 1915): 230–32.

27. Comment from Miss Minnie H. Ahrens (Chicago) during a discussion following Miss Hyde's paper on private-duty nursing at the 1906 private-duty session of the nursing meetings, ANA Collection, box 34, folder 9, Mugar Library, Boston University.

28. This problem of not being able to distinguish between organizations is discussed in Grace Cook, "The Value of a Local Directory," *AJN* 22 (September 1922): 1021–23.

29. Gopsill's *Philadelphia Business Directory* (Philadelphia: James Gopsill, 1884) contained no advertisements, only a listing of names and addresses. The Philadelphia Directory, during its fifty-two years of existence (1882–1935), contributed $97,805 to Philadelphia's College of Physicians, and 150,000 calls were filled. Nurses did not form their own directory in Philadelphia until 1929. See "Report of Committee on Directory for Nurses," May 1935, CPP Collection, CPP 10/0001-01. See also Roberta Mayhew West, *History of Nursing in Pennsylvania* (Pennsylvania State Nurses' Association, 1939), 34–35. Boston listed nurses only by name and address; see, for example, *The Boston Directory* (Boston: Sampson, Mordock, 1901), 1972–74.

30. Jane Edna Hunter, *A Nickel and a Prayer* (Cleveland: Elli Kani Publishing, 1940), 52–77. In Cleveland, "the question of colored members" was first discussed in 1908; see Minutes of the Registry Committee, 31 March 1908, GCNA Collection. The rejected application incident was reported in Minutes of the Registry Committee, 12 January 1922, and "History of the Registry," GCNA Collection. See also *Charleston City Directory* (Charleston, S.C.: Lucas & Richardson, 1895); the introduction to the directory reminded readers that "to find a name you should know how to spell it."

31. *Charleston City Directory;* "History of the Registry," 1: 3. See also Minutes of the Registry Committee, 29 May 1906 and 5 April 1921, GCNA Collection; Report of the Committee on Directory for Nurses, 1 March 1882, CPP Collection, CPP 10/0001-01; "Regulations for the Central Registry for Nurses," 19 March 1926, and "14th Annual Report of the Suffolk Co. Nurses Central Directory," 19 March 1926, MNA Collection.

For an overview of the impact of the telephone, see Emily Abel, "'A Terrible and Exhausting Struggle': Family Caregiving during the Transformation of Medicine," *Journal of the History of Medicine and Allied Sciences* 50 (October 1995): 478–506; and Ithiel de Sola Pool, ed., *The Social Impact of the Telephone* (Cambridge: MIT Press, 1977).

32. A sample survey of calls received by the registry run by the Alumnae Association of the Presbyterian Hospital in Philadelphia, June 1904, found that forty-four requests for nurses were filled, of which twenty-two went to nurses with phones. Of those with phones, fourteen lived in group houses. Six more jobs went to nurses living close to the hospital. See Presbyterian Hospital Collection, Center for the Study of the History of Nursing, University of Pennsylvania School of Nursing. See also Frances Taylor, "Private Nurses Need an Organization," *AJN* 16 (June 1916): 851–54. By 1917, it was said to be "absolutely necessary for a nurse to have a good telephone service"; see DeWitt, *Private-Duty Nursing,* 40.

33. de Sola Pool, *Social Impact of the Telephone,* 4.

34. Abel, "'Terrible and Exhausting Struggle'"; Registry Committee Minutes, 29 May 1906 and 11 April 1927, "History of the Registry," 1: 3, GCNA Collection.

35. L. L. Dock, "Nurses, Attention," *Trained Nurse* 26 (March 1901): 162; Holmes, "Ideal Central Directory." A book published by Miss L. E. Gordan of Philadelphia, listing nurses available through her registry, was sent to all the city's physicians and druggists; see A Well Wisher, "A New Directory," *Trained Nurse* 14 (January 1895): 98. Families disorganized by sickness are discussed in a paper presented at the Graduate Nurses Association of Virginia Convention in 1916, Graduate Nurses Association of Virginia Collection, folder 11. See also Wills, "Problem of Private Nursing," 79–80. The fear of "strangers" was discussed at the 1915 Graduate Nurses Association of Virginia Convention; see Holleman, "Private-Duty Nurse," 42–47.

36. Linna Richardson, "Reasons for Central Registries and Club Houses," *AJN* 9 (September 1909): 983–86.

37. Katherine DeWitt, "Specialities in Nursing," *AJN* 1 (October 1900): 14–17; Burgess, *Nurses, Patients, and Pocketbooks,* 71–76; Marion Mead, "Registry System of the Hennepin County Graduate Nurses' Association, Minneapolis, Minn.," *AJN* 10 (July 1910): 819–23; Keen et al., "To the Fellows"; Dear Sir from Registrar, n.d., Nurses Central Directory, Norfolk, Va., Graduate Nurses Association of Virginia Collection; Grace Cook, "Central Directories and Their Relation to Private-Duty Nurses," *AJN* 21 (December 1920): 148–51; Bertha Love, "Responsibility of a Successful Private-Duty Nurse to a Registry," *AJN* 16 (June 1916): 881–82; "History of the Registry," Report of the Registry Committee, 31 December 1907, and "Registry Record," 1: 3, GCNA Collection; Jane Mottus, *New York Nightingales: The Emergence of the Nursing Profession at Bellevue and New York Hospital, 1850–1920* (Ann Arbor, Mich.: UMI Research Press, 1981), 149–63.

New York Hospital's Graduate Nurse Club's Registry listed 227 nurses between 1898 and 1902, of which all but fifty listed some exceptions—cases they would not take; see New York Hospital Archives. See also Sara Parsons, "Report of the Central Directory, 1924," MNA Collection; Annual Reports to Registry Committee, 1905–1927, Collection of Central Registry Committee, GCNA Collection.

38. See, for example, "To Physicians," a flyer published by the Directory for Nurses of Washington, D.C., 15 December 1882, CPP Collection. See also "Report of Special Registry Committee," *AJN* 15 (September 1915): 1121–22; "The Commercial Directory," *AJN* 14 (February 1914): 329–30; Cook, "Central Directories"; Emma E. Collins, "What the Registry Means to the Private-Duty Nurse," *AJN* 28 (July 1928): 677–80.

39. Reverby, *Ordered to Care*, 95–117; Reverby, "Private-Duty Nursing in Boston"; U.S. Bureau of the Census, *Historical Statistics of the United States* (Washington, D.C.: U.S. Government Printing Office, 1959), Ser. B, 192–94; Burgess, *Nurses, Patients, and Pocketbooks*," 330; New York Hospital, Graduate Nurses' Club, "Schedule of Rates," June 1929, New York Hospital Archives.

40. Miss Mellichampe, "The Development and Value of a Nurses' Registry" (paper presented at the 1915 Graduate Nurses Association of Virginia Convention), 41–42, VNAR Collection. At this time the registry was eight years old.

41. Mellichampe, "Development and Value"; Mead, "Registry System," 819–22; "Regulations for the Central Registry for Nurses" (flyer for the Washington, D.C., Registry, n.d.), CPP Collection, CPP 10/0001-01; Holmes, "Ideal Central Directory," 606–8; Elvira Neubauer, "The Private Duty Nurse and the Registry," *AJN* 24 (January 1924): 277–80.

42. It was not unheard of for registrars to be considered "unfit" and dismissed. See, for example, Claudia Wheeler, Chairman, Registry Committee, to the Board of Trustees of District No. 4 of Ohio State Association of Graduate Nurses, August 1919, GCNA Collection. The problem of stealing is mentioned in "From the Brooklyn Eagle Newspaper," *Trained Nurse* 9 (January 1894): 194. Brigham to Dear Dr. "Scrapbook" entries for 1898 and 1902 outline confidential questions asked and rules for the nurses. In 1906, the registry decided to ask all applicants for a photograph (6 March 1906). Also see "Regulations for the Central Registry," n.d., CPP Collection; Report, Registry Committee, 11 April 1927, GCNA Collection; S. Virginia Levis, "Side Lights on Private Nursing," *Trained Nurse* 37 (November 1906): 376–78; Judith Howard, "The Business Side of Nursing," *Trained Nurse* 30 (March 1903): 151–54; "F. E. Goodban to Editor," *Trained Nurse* 13 (October 1894): 367; Helena McMillan, "The Objects of the Graduate Nurses' Association of Cleveland," *AJN* 1 (October 1900): 187–94; Dr. Musser, "Hints for Private Nurses," *Trained Nurse* 4 (March 1890): 126–27.

43. See all sources cited in n. 42.

44. "Goodban to Editor." Additionally, registries had rules outlining the nurses' duties to the registry, including availability, rates, and payment of registry fees. A first failure to adhere to the rules customarily resulted in a printed warning; second offenses ensured suspension of registry privileges. Competency was seemingly less important than ability to fit into the patient's household or to follow the registry rules.

45. The Grouch, "Nursing Problems," *Trained Nurse* 55 (August 1915): 110; A.E.C., "In Answer to the Grouch," *Trained Nurse* 55 (October 1915): 250.

46. Geister, "Private-Duty Nursing"; L. L. Dock, *Short Papers on Nursing Subjects* (New

York: M. Louise Longeway, 1900), 40–41; DeWitt, *Private-Duty Nursing,* 16–17. See also "The Patient's Family (Editorial)," *Trained Nurse* 46 (1911): 166; Thomas E. Satterwaite, M.D., "Private Nurses and Nursing, with Recommendations for Their Betterment," *Trained Nurse* 44 (April 1910): 211–16.

47. Geister, "Private-Duty Nursing"; Geister, "Hearsay and Fact in Private Duty," *AJN* 26 (July 1926): 515–27; Reverby, "'Something besides Waiting,'" 133–56; Burgess, *Nurses, Patients, and Pocketbooks.* For physicians' part in these changes, see George Rosen, ed., *The Structure of American Medical Practice* (Philadelphia: University of Pennsylvania Press, 1983); Edward Atwater, "The Physicians of Rochester, N.Y., 1860–1910," *Bulletin of the History of Medicine* 51 (1977): 93–106; Neil Larry Shumsky, James Bohland, and Paul Knox, "Separating Doctors' Homes and Doctors' Offices: San Francisco, 1881–1941," *Social Science and Medicine* 23 (1986): 1051–57.

The heading for this section is from Anne Williamson, *Fifty Years in Starch* (Culver City, Calif.: Murray and Gee, 1948).

48. C. Rufus Rorem, "Nursing—an Economic Paradox," in *Proceedings, 39th Annual Convention, National League of Nursing Education, 1933* (New York: National League of Nursing Education, 1933), 119. For registry records, see the following in the GCNA Collection: Registry Reports, 10 February 1920, 16 March 1922, 1 January 1924, June 1927, and September 1927; "History of the Registry," 1: 3; reports dated January 1918 and 22 January 1919, 4: 6; "Annual Report of the Registry," 1 October 1910, 4: 5; "Report of Calls," 1 October 1911, 4: 4; "Numbers of Calls," 18 January 1926, 4: 8. On overcrowding in the field, see An Observer, "Is the Profession Becoming Overcrowded?" *AJN* 3 (April 1903): 513–15; Burgess, *Nurses, Patients, and Pocketbooks,* 74–78.

49. Charles Rosenberg, *The Care of Strangers: The Rise of America's Hospital System* (New York: Basic Books, 1987), 341.

50. Ibid. And see, for example, "History of the Registry," 16 March 1922, April 1923, and 20 October 1926, GCNA Collection.

51. Sara Parsons, "Report of the Central Directory, 1924" and "14th Annual Report," Boston's Central Directory Collection, Mugar Library, Boston University. For Boston's directory, work was very seasonal and "nurses were suffering for work." Calls grew from 790 in 1912 to 10,990 by 1924. In 1924, 9,408 of these were from hospitals; calls came from twenty-eight hospitals and 370 physicians, but mostly from two hospitals and twenty-five to thirty physicians. During that year, only 1 percent of requests (1,530) were for home care. The first mention of a call from a hospital requesting private-duty nurses was in August 1915. The percentage of hospital cases quickly increased from half of all cases in 1917 to two-thirds by 1920. See "MNA Central Directory Records," 1912–1921, MNA Collection.

52. Several historians have examined this transformation. See Whelan, "Too Many, Too Few"; Reverby, "'Something besides Waiting,'" 133–56; Susan Reverby, "The Search for the Hospital Yardstick: Nursing and the Rationalization of Work," in Susan Reverby and David Rosner, eds., *Health Care in America: Essays in Social History* (Philadelphia: Temple University Press, 1979), 206–25. See also Marilyn Flood, "The Troubling Expedient: General Staff Nursing in United States Hospitals in the 1930s: A Means to Institutional, Educational, and Personal Ends" (Ph.D. diss., University of California, Berkeley,

1981); "What Registries Did in 1937," *AJN* 38 (October 1938): 1115–24; "Can You Send Me a Nurse?" *AJN* 38 (October 1938): 1113–15; Judith Whitaker, "Registries Then and Now," *AJN* 52 (October 1952): 1205; "Registries and Dollars," *AJN* 55 (April 1955): 444–45.

53. See, for example, Philadelphia's phone directory (published by the Bell Telephone Company of Pennsylvania) for the 1950s.

Chapter 7: A Cautionary Tale

1. Annual Report 1910, 19, DNAP Collection; Annie Brainard, "The Administrative Side of Visiting Nursing," *PHN* 8 (January 1915): 73–81; One of Them [Mrs. John Lowman], "Concerning a Few of the Duties and Privileges of Trustees," *VNQ* 4 (April 1912): 25. Gardner's comment is from Annual Report 1913, 30, DNAP Collection.

2. One of the Older Women, "A Decade of Change," *PHN* 5 (January 1913): 67. In 1913, Boston VNA's budget was $73,000, Philadelphia's $41,000. See Annual Report 1913, IDNA Collection; Annual Report 1913, VNSP Collection. Philadelphia experienced the appalling deficit in 1915, as did Providence and Boston. See Annual Report 1915, 4, VNSP Collection; Annual Report 1915, 7, DNAP Collection; Annual Reports 1911, 19, and 1913, 13, IDNA Collection.

3. "The Metropolitan Makes New Appointment: Dr. Frankel Becomes Manager of the Industrial Department," Lee Frankel files 1–2; Lillian Wald to Alma Haupt, 26 February 1937; Lee Frankel, Welfare Division, Nursing Bureau, file 160607; Lee Frankel, "Insurance Companies and Public Health Activities" (paper presented at the National Public Health Association, September 1913), 3; all in MLI Collection.

Industrial insurance was a new concept; see Alma Haupt, "Nursing and Life Insurance," *International Nursing Review* 11 (1937): 471–76. For a more extensive discussion of these origins, see Diane Hamilton, "Faith and Finance," *Image: Journal of Nursing Scholarship* 20 (fall 1988): 124–27; R. L. Duffus, *Lillian Wald: Neighbor and Crusader* (New York: Henry Holt, 1915), 1–5. The data Wald presented were probably from Jane Hitchcock, "Five Hundred Cases of Pneumonia," *AJN* 3 (December 1902): 169.

4. Frankel, "Insurance Companies," 6. Life insurance as a social institution remains an ongoing experience for the MLI. See, for example, "MetLife Plaintiffs Face a Decision," *Philadelphia Inquirer*, 24 October 1999, E1. This article reports on the outcome of a long-running class action dispute over the company's deceptive sales practices. See also Bruce Lewenstein, "Popular Science Information as a Tool: A Case Study in Public Health" (unpublished paper, 1984); Diane Hamilton, "The Metropolitan Life Insurance Company Visiting Nurse Service (1909–1953)" (Ph.D. diss., University of Virginia, 1987), 15–30; James Marquis, *The Metropolitan Life Company: A Study in Business Growth* (New York: Viking Press, 1951), 139–65.

5. Wald to Haupt, 26 February 1937; Alma Haupt, "Forty Years of Team Work in Public Health," *AJN* 53 (June 1953): 81–84; Alma Haupt, "The Metropolitan Story," *Quarterly Bulletin for Metropolitan Nurses* 15 (November 1951): 4.

6. Haupt, "Metropolitan Story."

7. See Haupt's "Metropolitan Story," "Nursing and Life Insurance," and "Forty Years of Team Work."

8. Haupt's "Metropolitan Story," "Nursing and Life Insurance," and "Forty Years of Team Work."

9. Lee K. Frankel and Louis I. Dublin, "Visiting Nursing and Life Insurance: A Statistical Summary of Results of Eight Years," *American Statistical Association Journal* 16 (June 1918): 58–60; MLI, "The Visiting Nurse Service: Conducted by the Metropolitan Life Insurance Company for the Benefit of Its Industrial Policyholders," 1917, MLI Collection.

10. "Shall the Visiting Nurse Continue on the Basis of Charity or Become a Factor in the Business Organization?" *Insurance and Investment News* (March 26, 1914): 1.

11. Lee Frankel, "Visiting Nursing from a Business Organization's Standpoint," *PHN* 5 (July 1913): 25–45.

12. By 1915, the arrangements for payment were based on "the cost of each call," that is, the nurse's salary divided by the average number of visits she made per month. Car fare and other incidental expenses, such as telephones, were also supplied. And "an additional charge is also made for the Company's share in the cost of supervisory, executive, and clerical staff." While it was understood that this was not an exact charge, it proved satisfactory. See Ella Crandall, "A New Extension of Visiting Nursing," *AJN* 10 (January 1910): 237–38; Frankel, "From a Business Organization's Standpoint."

13. Frankel, "From a Business Organization's Standpoint"; Crandall, "New Extension."

14. Isabel Lowman, "The Need of a Standard for Visiting Nursing," *VNQ* 4 (January 1912): 8–16. For a detailed description of the MLI situation, see Ella Crandall, "Memoranda on Circumstances Leading to the Organization of the NOPHN," May 1921, and Eleanor Mumford, "Field Interview with Ella Crandall," 19 January 1937, Gardner Collection, folder 45, Schlesinger Library; Grace Allison, "Shall Attendants Be Trained and Registered?" *AJN* 12 (August 1912): 933. In 1910 the MLI apparently supplied the VNSP with an attendant "to do necessary nursing for MLI Co. where the services of a trained nurse were not necessary"; see "Staff Superintendent Report, 1 July–1 August 1910," VNSP Collection.

15. Wald and Crandall discussed the matter of practical nurses with Frankel, reminding him of the terms and conditions stipulated by Wald when the service was initiated. To Wald and Crandall, the company's action seemed to indicate a serious tendency toward commercialism. Crandall, "Memoranda."

16. Edna Foley, "Concerning the Employment of Practical Nurses by Visiting Nurse Associations," *AJN* 12 (January 1912): 328, 330.

17. *The Welfare Work of the Metropolitan Life Insurance Company for Its Industrial Policyholders: Report for 1915* (New York: MLI, 1915), 3; Lowman, "Need of a Standard." Beard mentions that during the previous year the MLI had "cut down the number of visits to chronics to 25"; see "Report of the Work, June 1912," IDNA Collection.

18. Lowman, "Need of a Standard." For an extensive discussion of the NOPHN, see Louise Fitzpatrick, *The National Organization for Public Health Nursing, 1912–1952: Development of a Practice Field* (New York: National League for Nursing, 1975).

19. Frankel, "From a Business Organization's Standpoint"; Ella Crandall to Jane Delano, 3 January 1913; Miss Crandall, untitled circular to the readers of the *VNQ*, disapproving of Dr. Frankel's plan, Records of the American Red Cross, 1886–1916, National Archive Collection, RG 200, box 40, folder 500.002.

20. Crandall, untitled circular.

21. Crandall, "Memoranda." For a discussion of Frankel's speech, see Annual Report 1913, 28–29, DNAP Collection; Annual Report 1913, 17, IDNA Collection. Frankel's plan for a separate staff for MLI policyholders was finally laid to rest in a special meeting of the executive committee of the NOPHN, several lay members (VNA board members), and Frankel. See Minutes of the Executive Committee, 21 June 1913, NOPHN Collection. For Frankel's proposal to one VNA for a separate staff and its rejection of the idea, see Minutes of the Board, 13 December 1912, VNSP Collection. See also Minutes of the Board, 25 June 1915, NOPHN Collection.

22. "Nursing Service Is Not a Charity," MLI Collection.

23. Frankel was so eager to discuss his proposal with the members of the executive committee that he even offered to pay their expenses to New York at their earliest convenience. Crandall eventually convinced him to postpone this discussion until the nursing convention. See Ella Crandall to Jane Delano, 11 January 1913, Records of the American Red Cross, 1886–1916, National Archive Collection, RG 200, box 40, folder 500.002. Mary Beard's comment is in her "President's Address," *Proceedings of the Twenty-third Annual Convention, National League for Nursing Education, 1917,* 61.

24. This work with paying patients was also called "hourly nursing." See Annual Report 1916, VNSP Collection; Alma Wrigles, "The Hourly Nurse and Her Place," *AJN* 16 (April 1916): 874–81; Minutes, 25 June 1915, VNSP Collection. The Cleveland experiment with hourly nursing is discussed in Minutes of the Executive Committee Meeting of the NOPHN, 18–20 January 1915, roll 31, NYPL.

25. Frankel, "From a Business Organization's Standpoint"; Report of the Executive Committee of the NOPHN, 21–27 June 1913, Wald Collection, roll 31, NYPL. Frankel's concerns were next discussed at the 25 June 1915 board of directors meeting. Although "the secretary [Ella Crandall] was instructed to continue to urge, by all possible means, the establishment of a fee system as a matter of principle," little materially changed. See Minutes, Board of Directors Meeting, 25 June 1915, NOPHN Collection, roll 11.

26. Gardner's comment is in Annual Report 1913, 28–29, DNAP Collection. See also Ella Crandall, "Care of the Sick in Their Homes," *PHN* 7 (January 1915): 13.

27. Prenatal care was added as a benefit in 1916; see Frankel and Dublin, "Visiting Nursing and Life Insurance," 37. Initially, policies could not be purchased for children under one year of age, but by 1915 the mortality of this age group had declined sufficiently to make them an acceptable risk. See Louis Dublin, *The Effects of Life Conservation on the Mortality of Metropolitan Life Company: A Summary of the Experience, Industrial Department, 1914, for Agents, Medical Examiners, and Visiting Nurses* (New York: MLI, 1914), 4–8; Haupt, "Forty Years of Team Work"; Haupt, "Nursing and Life Insurance"; MLI, *Welfare Work of MLI, 1915* (in-house pamphlet). In 1918, the MLI added a new market by offering companies group policies. For a discussion of the types of problems created, see Karen Buhler-Wilkerson, "Caring in Its 'Proper Place': Race and Benevolence in Charleston, S.C., 1813–1930," *Nursing Research* 41 (January/February 1992): 14–20; see also "Essentials of Group Nursing: Suggestions for Nurses Interviewing Executives of Metropolitan Group Policyholders," n.d., MLI Collection, folder 160606.

28. These mortality data are for males fifteen and over. See Louis Dublin, Edwin Kopf,

and George Van Buren, "*Mortality Statistics of Insured Wage-Earners and Their Families: Experiences of the Metropolitan Life Insurance Company Industrial Department, 1911–1916, in the United States* (New York: MLI, 1919), 2–7; Louis Dublin and Alfred Lotka, *Twenty-five Years of Health Progress: A Study of the Mortality Experience among Industrial Policy-holders of the Metropolitan Life Insurance Company, 1911–1930* (New York: MLI, 1937), 7–10. By 1934, compared with the general population, policyholders' representation was greater in industries than in the professions and agriculture. The insured population also had a greater proportion of women and a younger average age than the general population.

29. Marquis James, *The Metropolitan Life: A Study in Business Growth* (New York: Viking Press, 1947), 86, 338–39; Walter Weare, *Black Business in the New South: A Social History of the North Carolina Mutual Life Insurance Company* (Urbana: University of Illinois Press, 1973), 98; Dublin, *Effects of Life Conservation;* Louis Dublin, "The Effect of Health Education on Negro Mortality," *Opportunity* 2 (August 1924): 232–34; Lee Frankel, "A Study of Mortality in Southern Communities" (paper presented at the Southern Sociological Congress, April 1916), MLI Collection; Louis Dublin, *After Eighty Years: The Impact of Life Insurance on Public Health* (Gainesville: University of Florida, 1966), 148–53.

Only 12.3 percent of policyholders were black, but they accounted for 16.9 percent of the claims in 1914. By 1923, there were 1.8 million black policyholders, and by 1928, 2.5 million. In 1928, the MLI had twenty times more insurance on blacks than did North Carolina Mutual. See Weare, *Black Business in the New South,* 130; Buhler-Wilkerson, "Caring in Its 'Proper Place.'" One supervisor reported that, though word was still unofficial, the MLI was planning to no longer write policies for colored people; see Board of Managers Meeting, 2 February 1928, VNSP Collection.

30. Dublin's conclusions, which appeared in an article titled "Life, Death, and the Negro," were published in the *American Mercury* and reported in "Statistical Bulletin, Metropolitan Life Insurance Co., August 1927," 8–9.

31. The heading for this section is from Dublin and Lotka, *Twenty-five Years,* 3. Maurice Taylor claimed that, on average, each policyholder carried 1.7 policies. Thus wherever the MLI gives the number of "policies in force," fewer individuals were actually involved. See Maurice Taylor, *The Social Cost of Industrial Insurance* (New York: Alfred Knopf, 1933), 284–85; *Welfare Work of MLI, 1915,* 6. By 1923, twenty-seven per thousand policies in force received care from the nursing service; see "An Appraisal of Insurance Visiting Nurse Work" (unpublished MLI study, 1923), 21.

32. Louis Dublin, "Records of Public Health Nursing: Tabulation and Analysis of Nursing Records in Visiting Nurse Work," *PHN* 13 (October 1921): 518–31. Dublin argued that this difference between percentage of black policyholders and percentage of blacks in the general population was due to "choice of localities in which the company conducts industrial life insurance business." See Dublin et al., *Mortality Statistics,* 5; Louis Dublin, "Records of Public Health Nursing and Their Service in Case Work, Administration, and Research," *PHN* 13 (June 1921): 385–92.

33. The MLI concluded that "had the 1911 rate continued into 1915, the number of deaths would have been 11,800 more than were actually registered." Most impressive were the percentage declines in MLI death rates for typhoid fever, pneumonia, infectious dis-

eases of childhood, tuberculosis, puerperal diseases, and accidents. Frankel and Dublin, "Visiting Nursing and Life Insurance," 53–54.

34. Lee K. Frankel, "Standards in Visiting Nurse Work," *AJN* 15 (August 1915): 984–95. Recognizing that the variability among agencies suggested the need for standard nomenclature, Frankel urged the standardizing and classification of terms to allow a more accurate analysis among VNAs. Crandall's comment was included in the discussion following Frankel's paper; ibid., 994–95. By the 1916 NOPHN convention, a committee headed by Frankel had developed a standard record card to be adopted by all VNAs; see "The Fourth Annual Report of the NOPHN," *PHN* 8 (July 1916): 11–12, 17. It took the NOPHN another year to appoint a committee to consider the need for a standardized financial statement. By 1919, that had been accomplished. See "Report of the Committee on Organization and Administration," 31 December 1917 and 14 June 1918, NOPHN Collection, roll 25. The MLI series of articles included, for example, Lee Frankel, "Standardization of Financial Statements for Visiting Nurse Associations," *PHN* 8 (July 1916): 67–71; Dublin, "Records in Case Work"; and Dublin, "Records: Tabulation and Analysis."

35. Frankel, "Standards." See also memo accompanying the "Visiting Nurse Study, 1918," prepared by the MLI statistical division, and E.B. to Frankel, 24 May 1917, Welfare Division Folder, Nursing Service, 1909–1922, MLI Collection.

36. Frankel, "Standards."

37. Frankel and Dublin, "Visiting Nursing and Life Insurance," 40.

38. Ibid., 39. Practical case management meant simply the proper discharge of cases when the services of the nurse were no longer needed and the proper choice of cases at the proper stage of illness. Dublin, *Effects of Life Conservation,* 3–11.

39. Frankel and Dublin, "Visiting Nursing and Life Insurance," 11.

40. "Appraisal of Insurance." For examples of number of visits per case, see annual reports of VNAs.

41. Frankel and Dublin, "Visiting Nursing and Life Insurance," 11, 60. In 1922, the MLI added eight newborn visits. See Dublin, *Effects of Life Conservation,* 3–11, 37; Lee Frankel, *Welfare Work of the Metropolitan Life Insurance Company for Its Industrial Policyholders: Report for 1918* (New York: MLI, 1918). For number of visits per case, see annual reports for Boston and Philadelphia, 1900–1910. See also Taylor, *Social Cost.* The increase in maternity cases at least partly reflects the administrative decision made in 1914 to extend the maternity service from six to eleven prenatal visits in addition to the traditional eight postnatal visits. In 1916, 26 percent of all female policyholders receiving care were maternity patients. With 19.5 maternity cases per thousand female policyholders using the service in 1916, compared with 15.8 per thousand in 1914, the MLI could claim to have succeeded in extending the work of its visiting nurse service to an increasing number of women.

42. Frankel and Dublin, "Visiting Nursing and Life Insurance," 13, 51–52, 60; "Statistical Report for 1931," covering August 1930–July 1931, MLI Collection; "Appraisal of Insurance," 13, 28–33. Each case of tuberculosis, for example, cost $6.56 in 1914, while a cancer case cost $8.35. The only really expensive acute conditions were typhoid, puerperal septicemia, and burns. The high cost of caring for all three reflected the greater number of visits required (14, 12.8, and 12.3 per case, respectively).

43. "Appraisal of Insurance," 5–18; "Statistical Report for 1931."

44. "Appraisal of Insurance," 29.

45. Ibid., 20–21, is the first record of this concern about care in maternity cases. Dublin discussed his plans with Wald and the Nurse Committee at Henry Street Settlement in 1928. The conversation was described as animated. Minutes, Nurse Committee, 23 March 1928, VNSNY Collection.

46. Dublin, *Effects of Life Conservation*, 7–8; C. C. Spaulding, "Improvement in Negro Health as Shown by Insurance Research," *Opportunity* 1 (December 1923): 365; Taylor, *Social Cost*, 293–94.

47. "Stabilization of Metropolitan Visiting Nurse Service," 1929, MLI Collection.

48. Dublin, "Records: Administration and Research."

49. Lee Frankel to Louis Dublin, 19 September 1924, MLI Collection.

50. "Appraisal of Insurance." The MLI admitted that its ability to classify quality accurately depended on the quality of case records and that such a procedure was "fraught with great danger of error."

51. "Appraisal of Insurance," 33.

52. These data on denial of payments are difficult to find. A 1920 study published in *PHN* found the following percentages of visits not paid for: Philadelphia, 14 percent; Providence 18 percent; Washington, D.C., 32 percent; Richmond, 28 percent; and Boston, 35 percent. An apology for not including the classification of these cases was later published, with a regret that these data might have been "misleading." See "Abbreviated Three-Month Study of Visits Made to Metropolitan Life Insurance Company Patients Not Paid for by the Company," *PHN* (July 1920): 616; "MLI Company Nursing Visits," *PHN* (December 1920): 1020. See also "Home Office Analysis of Nursing Service, 1931" and "Stabilization," 1929, MLI Collection. For a more extensive examination of Frankel's conclusions, see Hamilton, "MLI Visiting Nurse Service," 137–41.

53. The MLI grant to the NOPHN to conduct this study is discussed in "Resolution Adopted at the Business Meeting, NOPHN, 1 July 1922," Report of the Executive Committee, 18–19 September 1922, NOPHN Collection, roll 11. The study was completed in 1924. See Minutes of the Board of Directors, NOPHN Meeting, 21 June 1924, NOPHN Collection; *Report of the Committee to Study Visiting Nursing, Instituted by the NOPHN at the Request of the Metropolitan Life Insurance Company* (New York: MLI, 1924). See also Minutes of the Executive Committee, NOPHN, 12–16 February 1922, 18–19 September 1922, and 30 October to 1 November 1922, NOPHN Collection, roll 11; Minutes of the Executive Committee, NOPHN, 10–11 March 1925, NOPHN Collection, roll 11; "Report of the Conference of Representatives of Voluntary (Non-official, Non-governmental) Private Public Health Associations," 7 May 1925, Wald Collection, roll 31, NYPL; Hamilton, "Faith and Finance," 124–27. A later study reported that while insurance companies were charged "cost" per visit, individual charges were commonly higher than actual cost. See NOPHN, *Survey of Public Health Nursing: Administration and Practice* (New York: Commonwealth Fund, 1934), 121–26, NYPL Collection, roll 31; Nurse Committee, 11 May 1925, CHA, IDNA Collection.

54. Louis Dublin to Lee Frankel, "Philadelphia Visiting Nursing—Metropolitan and Non-Metropolitan Cases Compared," 1 December 1924, MLI Collection.

55. Taylor, *Social Cost*, 276, 283–87; Lee Frankel, "The Life Insurance Company in

Health Conservation Programs," *Annals of the American Academy of Political and Social Science* (1925): 1–8; "What Did We Do in 1921?" MLI Collection; Frankel and Dublin, "Visiting Nursing and Life Insurance," 55.

56. "Stabilization"; Lee Frankel, "Life Insurance and Public Health" (address presented at the International Convention of Life Underwriters, September 1926), MLI Collection; Diane Hamilton, "The Cost of Caring: The Metropolitan Life Insurance Company's Visiting Nurse Service, 1909–1953," *Bulletin of the History of Medicine* 63 (fall 1989): 414–34.

57. "MLI Statistical Bulletin, March 1929," quoted in Taylor, *Social Cost*, 291–92.

58. The heading of this section is from Taylor, *Social Cost.*

59. Ibid.

60. The details of these changes are well documented in Hamilton, "Cost of Caring."

61. The concern of VNAs about declines in insurance cases was blamed on a variety of causes—increase in chronic illness, six-month waiting time before new policyholders could receive care, and increasing access of policyholders to hospitals. See, for example, Board of Managers Minutes, January 1944 and May 1945, IDNA Collection. For a discussion of the MLI's worries about cost rising faster than volume of service, see Board of Managers Minutes, April 1948, IDNA Collection. See also Alma Haupt, "1909–1952, The Metropolitan Nursing Service" (memo announcing termination, July 1950), Welfare Division, Nursing Service, MLI Collection; "Mission Accomplished," *Quarterly Bulletin for Metropolitan Nurses* 15 (November 1951).

62. Advertising Research Bureau, "Survey of Metropolitan Nursing Service," 1947, MLI Collection.

63. Haupt, "Forty Years of Team Work"; Ruth Fisher and Alma Haupt, "Community Effort to Provide Visiting Nurse Service: A Report of a Study of the Effect of Terminating the Metropolitan Nursing Service," *Nursing Outlook* 1 (January 1953): 46–48; Haupt, "1909–1952." See also Hamilton's "Faith and Finance" and "Cost of Caring." John Hancock Mutual Life Insurance and Aetna, both of which initiated visiting nurse services in the mid-1920s, also decided to discontinue these services. Leroy Lincoln to MLI field force, 30 June 1950, MLI Collection. See the following articles in *PHN* 43 (May 1951): "A Report of a Study of Effects of the Termination of Metropolitan Nursing Contracts," 285–89; Ruth Hubbard and Kathryn Frankenfield, "Comments on the Study: Part I, the Visiting Nurse Society of Philadelphia," 289–90; Mabel Reid, "Comments on the Study: Part II, the Visiting Nurse Service of New York," 290–92; and W. T. McCullough, "Comments on the Study: Part III, Community Chest of Philadelphia and Vicinity," 292–93.

Chapter 8: *"An Unchanging Purpose in a Changing World"*

The title of this chapter is from a plaque presented to the Boston VNA, November 1936.

1. Mary Beard's comment is in Annual Report 1919, 17, IDNA Collection. In Boston in 1917, each nurse cared for 296 patients, making 9.4 visits per patient; in 1918, each cared for 316 patients, making 7.9 visits per patient. Despite increased income, the IDNA's deficit remained stable at $2,000 per year. See Annual Reports 1917, 23, and 1918, 12, 39, IDNA Collection. In Providence, the number of patients cared for increased from 8,390

in 1917 to 10,332 in 1918, while the number of visits per patient decreased from 12.7 to 10.2. For a discussion of the flu epidemic, see Alfred W. Crosby, *America's Forgotten Pandemic: The Influenza of 1918* (Cambridge: Cambridge University Press, 1989). The VNSP cared for 4,050 patients during the epidemic, between 14 September and 2 November 1918; see Minutes of the Managers Meeting, November 1918, VNSP Collection. See also Annual Report 1918, 13, IDNA Collection. The work of Providence's DNAP during the epidemic is discussed in Annual Report 1918, 17–19, DNAP Collection. See also Annual Report 1919, 12–13, IDNA Collection; Fitzhugh Mullan, *Plagues and Politics: The Story of the United States Public Health Service* (New York: Basic Books, 1989), 74–75; Annual Reports 1917, 15; 1918, 17; and 1919, 8—all in DNAP Collection; Bessie Amerman Harris, "A Wartime Convention," *PHN* 10 (July 1918): 253–56.

2. In Boston, while the IDNA's staff increased by 25 percent, its budgets increased 43 percent, from $151,000 in 1919 to $216,000 in 1920. In 1919, patients' fees and the MLI contributed 23 percent of the IDNA's income; by 1920, this had increased to 29 percent. Boston ended the year with an $8,100 deficit. See Annual Reports 1919, 35–40, and 1920, 31–32, IDNA Collection. See also Mary Beard, "A Review," September 1921, 3, IDNA Collection. In Philadelphia, the VNSP added twenty-three nurses (35% increase) to its staff and $25,000 to its budget between 1920 and 1921; see Annual Reports 1920, 8, and 1921, 11, VNSP Collection. In Providence, the DNAP had a comparable 37 percent increase in staff but a smaller increase in budget, in part because of its low salary scale; see Annual Reports 1920, 21, and 1922, 21, DNAP Collection. Of all the organizations surveyed for the January 1920 NOPHN meeting, Providence paid the lowest salaries. See "Nurse Committee Notes," 1919, VNSP Collection.

3. For Philadelphia, see Board of Managers Meetings, 16 January 1925, 6 March 1925, and 13 March 1925; Board of Managers Meetings 13, 20, 24, 25 January 1922, 24 February 1922, 28 April 1922, and 23 January 1925; all in VNSP Collection. The salary paid in twelve larger VNAs in 1924 and the analysis of staff living expenses for Philadelphia in 1924 are in Nurse Committee Report, 1924, VNSP Collection. For Boston, see CHA, Nurse Committee, 12 November 1923, and Board of Managers Meeting, 21 December 1923, IDNA Collection.

4. Louis Dublin's claims about a "precarious situation" are in his "Records of Public Health Nursing," Lecture V, *PHN* 14 (January 1922): 17–24. His claims are substantiated by the annual reports of the IDNA, VNSP, Henry Street Settlement, and Providence VNA. See Henry Street Nursing Settlement, Nurse Committee Minutes, 13 January 1922 and 31 March 1922, VNSNY.

5. For an example of these financial problems at Henry Street, see Henry Street Nursing Settlement, Minutes of the Nurse Committee, 14 October 1921, 31 March 1922, 3 November 1922, 25 April 1924, 26 February 1926, 24 April 1926, 22 October 1926, 23 March 1928, and 22 November 1929, VNSNY Collection.

6. The only new program introduced by the VNAs during this period was in mental hygiene; compared with their recent projects, it was very limited. See, for example, Mary Gardner, *Public Health Nursing* (New York: Macmillan, 1924), 343–59.

7. MLI, "Appraisal of Insurance Visiting Nurse Work" (unpublished study as supplement to NOPHN's visiting nurse study, 1923), MLI Collection; Mary Beard, *The Nurse in Public Health* (New York: Harper and Brothers, 1929), 60–61.

8. Annual Reports 1919, 35–44; 1920, 31; 1922, 213–33; 1923, 12; and 1924, 8–10; all in IDNA Collection. Gardner discusses traditional methods of raising funds in "What Are Voluntary Organizations Going to Do?" 458; and Executive Committee Meetings, 20 January 1920 and 4, 10, and 31 March 1920, IDNA Collection. The Henry Street Settlement tried similar strategies. Beginning in the 1920s, it resorted to fundraising campaigns every two years. In 1926, it hired the firm of Tamblyn and Brown to launch one really big last campaign in the hope of completing its endowment. The goal was to raise $250,000 per year for two years, at an estimated expense of $20,000. The plan was undertaken in anticipation of a $50,000 budget overrun. See Henry Street Nursing Settlement, Minutes of Directors Meetings, 27 July 1926, 30 September 1926, and 21 March 1927, and "Program for the Financial Support of the Visiting Nurse Service," VNSNY Collection.

9. Executive Committee Meetings, 12 March 1920, 30 July 1920, and 10 December 1920, IDNA Collection. These fundraising methods were apparently considered successful enough for Miss Gertrude Peabody, vice president of the IDNA's board of managers, to publish them in her "Opportunities and Responsibilities for Lay Persons in Public Health Nursing Work," *PHN* 16 (May 1924): 229–34.

10. Nurse Committee Meeting, 29 February 1925, IDNA Collection. Local committees were mildly successful in fundraising. See also Executive Committee Meeting, 12 March 1920, 30 July 1920, and 10 December 1920, IDNA Collection.

11. For Beard's first report to the board on limiting work, see CHA, Minutes of Executive Committee Meeting, 28 February 1923, IDNA Collection. Beard discussed her report with the supervisors at their regular meeting; see CHA, Minutes of Supervisors Meeting, 21 January 1923, IDNA Collection. IDNA's financial difficulties had apparently been heightened after merger with the Baby Hygiene Association to form the CHA; the CHA took on all the child welfare work in the city, and its staff increased from 123 to 161 nurses. See CHA, Annual Report 1922, IDNA Collection; Mary Beard, "The Community Health Association of Boston: An Outline of the Plan of Organization Following Combining the Work of the Instructive District Nurse Association and the Baby Hygiene Association," *PHN* 15 (March 1923): 115–18. For the November 1923 meeting, see CHA, Minutes of the Supervisors Meeting, 12 November 1923, IDNA Collection.

12. The question of eliminating work in anticipation of staff reductions was first discussed at the 1 April 1924 supervisors' meeting. At the 16 April 1924 supervisors' meeting, Mary Beard's frustration was apparent when she described the staff's continued cooperation as "remarkable." Amazingly, not until September 1923 did Beard finally succeed in implementing to her full satisfaction the plans she had initiated in 1915. See CHA, Minutes of the Supervisors Meetings, 1 April 1924 and 16 April 1924, and Minutes of the Executive Committee, 19 September 1923, IDNA Collection. Thus, the plans had been in place for only seven months when the board decided to limit the work.

13. CHA, Minutes of the Nurse Committee, 14 April 1924, and Minutes of Supervisors Meetings, 16 April 1924, 23 May 1924, and 1 July 1924; CHA, Minutes of the Nurse Committee, 27 June 1923, 6 August 1923, 8 September 1923, 12 November 1923, and 14 May 1924; Mary Beard to David Edsal, 10 January 1921; all in IDNA Collection.

14. CHA, Minutes of the Board, 10 October 1924; CHA, Annual Reports 1924, 3–8, and 1925, 16; all in IDNA Collection. Florence Patterson became director of the CHA when Mary Beard resigned in October 1924.

15. The main focus of the convention was community demand for expanded services and the related need for increased funding. See, for example, "High Visibility in the Convention Program," *PHN* 16 (April 1924): 167–70. Many of the papers presented at the convention were published in the September and October 1924 issues of *PHN*. Of the five thousand people attending the combined convention (National League for Nursing Education, ANA, and NOPHN), 957 were members of the NOPHN; of these, 883 were nurses and the remainder were lay members. This registration was double that of past years. See "Activities of NOPHN," *PHN* 16 (September 1924): 363.

Starting with the June 1924 *PHN*, a new series was begun to offer the "flappers" of public health nursing the opportunity to "display their latest and most extreme innovations in management" and to explain the "broad and lasting results achieved on a small investment." It also offered the conservative an equal opportunity to defend "her low heels, steady gait, and reverence for tradition." See Florence Patterson, "A New Department and a New Series," *PHN* 16 (June 1924): 273–74.

16. William Norton, "Meeting the Demands for Community Health Work," *PHN* 16 (September 1924): 490–93. Three papers of this title were given at the convention, but Norton's was the focus of the roundtable discussion chaired by Mary Gardner. The other papers were much more general and addressed the issues of educating the public and community demand for care. On the basis of Gardner's report of the discussion, we can assume that Norton's paper most directly addressed the issues most critical to the participants. The other papers with the same title were by Haven Emerson (*PHN* 16 [September 1924]: 485–89) and Ella Crandall (*PHN* 16 [October 1924]: 506–12).

17. See, for example, Eleanor Marsh, "Telling the Public about the Public Health Nurse," *PHN* 12 (March 1920): 218–22; Sherman Kingsley, "Principles and Methods of Money Raising," *PHN* 12 (June 1920): 511–19; Estelle Hunter, "The Essentials of Office Management," *PHN* 12 (May 1920): 397–405; Sophie Nelson, "The Question of Community Funds," *PHN* 16 (March 1924): 124–26; and Anna Behr, "Publicity, An Essential Part of the Public Health Nursing Program," *PHN* 16 (January 1924): 21–26.

18. This notion about Boston's problems being related to lack of a community chest (precursor to today's United Way) was reported in Minutes of Managers Meeting, 31 October 1924, VNSP Collection. The Philadelphia board concluded that initiation of a community fund drive in Boston would be inevitable, noting the CHA's $80,000 deficit as sufficient evidence. Philadelphia had by that time formed a Welfare Federation, and in 1920 had 439 agencies doing some kind of "social work," with aggregate budgets of $19 million to $20 million. See Kingsley, "Principles and Methods," 512. Gardner's comment is in her "What Are Voluntary Organizations Going to Do about Meeting Their Demands with the Funds Available?" *PHN* 16 (September 1924): 457–58.

19. Gardner, "What Are Voluntary Organizations Going to Do?" This was the general agreement reached by those attending the general session of the NOPHN convention, which followed the session "Meeting the Demands for Community Health Work."

20. Gardner, "What Are Voluntary Organization Going to Do?" 458. In fact, the number of new public health nursing associations opening each year had been declining since 1920, which was a peak year with the formation of 441 new associations; by 1921 this number had declined to 301, and by 1923 to 238. See Louise Tattershall, "Census of Public Health Nursing in the United States," *PHN* 18 (May 1926): 264.

Self-analysis in terms of end results was not a new concept for visiting nursing. Lee Frankel first introduced the idea in his 1916 paper on the need for standardized financial statements, which would allow VNAs to make comparisons from year to year and among organizations and thus evaluate their successes and failures. Lee Frankel, "Standardization of Financial Statements for Visiting Nurse Associations," *PHN* 8 (July 1916): 67–71.

21. The first substantial analysis of nursing was the Goldmark study, funded by the Rockefeller Foundation: Josephine Goldmark, *Nursing and Nursing Education in the United States* (New York: Macmillan, 1923). For a summary of the significance of these conclusions for public health nursing, see Mary Beard, "Discussion of the Rockefeller Report: Analysis of the Situation in the Public Health Field," *Proceedings of the 29th Annual Convention of the National League for Nursing Education, 1923*, 179–84. Lack of any significant reform of nursing education was later documented by May Burgess in *Nurses, Patients, and Pocketbooks* (New York: Committee on the Grading of Nursing Schools, 1928).

The second major study was supported by the MLI: *Report of the Committee to Study Visiting Nursing Instituted by the NOPHN at the Request of the Metropolitan Life Insurance Company* (New York: MLI, 1924). See also "Report of Round Table on Visiting Nursing Study," 20 June 1924, Winslow Collection, 87: 1416; Minutes of the Executive Committee, 18–19 September 1922 and 30 October–1 November 1922, NOPHN Collection, roll 11. The MLI also conducted its own study, "Appraisal," completed in 1923 but never published. See also Minutes of the Board of Managers, 6 June 1924, VNSP Collection.

For examples of further studies, see Marguerite Wales, "The Value of Measuring Rods in a Visiting Nurse Service," *PHN* 19 (March 1927): 117–21; Marguerite Wales, "Time and Cost Study Problems," *PHN* 20 (March 1928): 138–40; Emma Winslow, "Service Norms and Their Variation," *PHN* 22 (February 1930): 68–73 and (March 1930): 151–52; Emma Winslow, "The Measurement of Nurse Power," *PHN* 19 (September 1927): 492–98; Emilie Sargent, "More about Measuring Rods," *PHN* 19 (May 1927): 231–33; Ira Hiscock, "The Value of Records and the Annual Report," *PHN* 19 (March 1927): 167–71; Emma Winslow, "Service Costs and Program Planning," *PHN* 20 (November 1928): 569–74; W. F. Walker, "The Appraisal of Nursing Service," *PHN* 20 (October 1928): 518–24; Emma Winslow, "More about the Measurement of Nurse Power," *PHN* 20 (February 1928): 63–68.

22. See all sources cited in n. 21. The critical analysis by Susa Moore, associate editor of *Nation's Health*, was in response to the findings of the Committee to Study Visiting Nursing; see her "The Report of Visiting Nursing," *Nation's Health* (January 1925): 6. This was a much-diluted version of her views, the result of numerous letters exchanged with William Snow, who chaired the Committee to Study Visiting Nursing. See Susa Moore to William Snow, 5 January 1925, 15 January 1925, 29 January 1925; William Snow to Susa Moore, 10 January 1925, 22 January 1925, 7 March 1925; William Snow to C.-E. A. Winslow, 7 March 1925; and C.-E. A. Winslow to William Snow, 16 March 1925; all in Winslow Collection 88: 1423

23. Lee Frankel financed this small regional conference of VNA leaders. Its purpose was to discuss trends in public health work and to begin planning for the next ten years. Frankel thought nurses must consider the underlying reasons for their present activities as well as the basis for "the division of responsibility for community health work between public health nursing associations and other community health agencies, official and

non-official." See "Report of Conference of Representatives of Voluntary (Non-official, Non-governmental) Private Public Health Associations," 7 May 1925, Wald Collection, roll 13, NYPL. Florence Patterson also reported on this meeting in CHA, Minutes of the Nurse Committee Meeting, 11 May 1925, IDNA Collection.

24. Mary Gardner, "Modern Problems in the Public Health Nursing Field," *PHN* 21 (August 1912): 413–16; Frances Bolton, "A Cleveland Opinion of the Community Chest Idea," *PHN* 4 (December 1922): 610–16; Sophie Nelson, "The Question of Community Funds," *PHN* 16 (March 1924): 124–26; Gertrude Hussey, "What of the Community Chest?" *PHN* 20 (June 1928): 291–93; Gardner, "What Are Voluntary Organizations Going to Do?" 442–51.

25. Wald is quoted in the centennial exhibit catalogue published by the VNSNY, "Healing at Home: Visiting Nurse Service of New York, 1893–1993" (New York: VNSNY, 1993). She also discusses the need for municipal support and the comparative cost of hospital versus home care in her *Windows on Henry Street* (New York: Macmillan, 1934), 81–82.

26. Rosemary Stevens, *In Sickness and in Wealth: American Hospitals in the Twentieth Century* (New York: Basic Books, 1989), 105–39; "Hospital Service in the United States," *JAMA* 94 (March 1930): 921–76.

27. Henry Hurd, "The Hospital as a Factor in Modern Society," *Modern Hospital* (September 1913): 33.

28. The "Report of the Committee to Study Visiting Nursing" produced a great deal of interest on the part of the organizations participating in the committee. As a result, they focused more intently on productivity, staff efficiency, and cost per service. For examples, see Minutes of the Board of Managers, 6 June 1924, and Minutes of the Nurse Committee, 11 November 1928, VNSP Collection; Annual Report 1925, DNAP Collection; and CHA, Minutes of the Supervisors Meeting, 10 July 1925, IDNA Collection. See also Dorothy Deming, "Business Methods in Visiting Nurse Associations," *PHN* 18 (September 1926): 490–94; Winslow, "Service Costs"; Wales, "Value of Measuring Rods," 118; Mrs. Richard Noye, "Financial Problems," *PHN* 19 (June 1927): 318–20.

By the late 1920s, directors of some of the larger VNAs acknowledged the impact of hospitalization on their caseloads. See, for example, Annual Report 1928, 9, VNSP Collection; CHA, Minutes of the Nurse Committee, 25 October 1929, 8 April 1929, 22 April 1929, and 27 May 1930, IDNA Collection.

29. The heading for this section is from "When You Need a Part-Time Nurse, Call Murray Hill 5-7231," Henry Street Visiting Nurse Service, 1939, VNSNY Collection. Charging for care was not a new concept for most VNAs—it had simply taken on a new level of significance. See Deming, "Business Methods"; Elizabeth Fox, "The Economics of Nursing," *AJN* 29 (September 1929): 1037–44. An alternative solution was attempted by the Manhattan Health Society; its prepaid nursing service lasted from 1922 to 1925. See Olive Husk, "A Pioneer Self-Support Health Service," *PHN* 17 (April 1925): 184; Olive Husk, "The Manhattan Health Society," *PHN* 16 (January 1924): 27–32; C.-E. A. Winslow, "The Nursing Problem," *New England Journal of Medicine* 200 (February 1929): 268; Michael Davis, "The Meaning of the Hourly Nursing Experiment in Chicago," *AJN* 33 (February 1933): 111–12; Miriam Ames, "Hourly Nursing: Report of an Eighteen-Month Experiment," *AJN* 33 (February 1933): 113–23.

For an excellent discussion of the private-duty nurse's problems, see Susan Reverby, "'Something besides Waiting': The Politics of Private-Duty Nursing before the Depression," in Ellen Lagemann, ed., *Nursing History: New Perspectives, New Possibilities* (New York: Teachers College, 1983), 133–56. See also Janet Geister, "Hearsay and Facts in Private Duty," *AJN* 26 (July 1926): 515–28. For a description of hourly nursing from the perspective of the private-duty nurse, see Alma Wrigles, "The Hourly Nurse and Her Place," *AJN* 16 (April 1916): 874–81. In a comment at the end of this paper, Mary Lent suggests that "rightly, part of [hourly work] belongs to the VNA." See also Burgess, *Nurses, Patients, and Pocketbooks,* 349–53.

While VNAs were opening hourly services, so were many registries. For example, in 1926 the CHA's Committee on Hourly Nursing decided to initiate an hourly service through a central registry in Boston; see Minutes of the Nurse Committee, 8 June 1926, IDNA Collection. In Philadelphia, not until 1930 did the Central Nurses Registry decide to consider offering an hourly service. In February, it asked the VNSP whether this would be encroaching on the society's work; see Minutes of the Nurse Committee, 21 February 1930, VNSP Collection.

Nationally, the forum for discussion was the Joint Committee on Distribution, which was cooperatively funded in 1928 by the NOPHN, ANA, and National League for Nursing Education. Through careful committee appointments, the NOPHN worked to ensure the achievement of the outcomes it desired. While publicly creating the image of a cooperative endeavor, in the end the ANA chose to take no significant steps to organize hourly nursing on behalf of private-duty nurses. The distribution of nurses within the community was, they claimed, the responsibility of the NOPHN. See Minutes of the Executive Committee, January 1930, NOPHN Collection, roll 11; Minutes of the Subcommittee on Hourly Appointment Service of the Committee on Distribution of Nursing Services, 17 September 1932, 11 June 1933, and 20 January 1934, NOPHN Collection, roll 26. The conclusions of this joint committee were published in "Hourly Appointment Nursing Service: Some Tentative Standards Prepared by the Joint Committee on Distribution of Nursing Service," *AJN* 31 (May 1931): 567–69.

30. Mary Gardner's patient classification system is in Annual Report 1914, 29, DNAP Collection. Paying patients accounted for only 6 to 14 percent of most budgets. See, for example, Annual Reports 1909, 20; 1913, 26; and 1914, 29—all in DNAP Collection; Annual Report 1907, 5, VNSP Collection; Annual Report 1917, 23, 27, IDNA Collection. Philadelphia began caring for pay patients as early as 1888; see Annual Report 1888, VNSP Collection.

31. Ames, "Hourly Nursing"; Davis, "Meaning of the Hourly Nursing Experiment in Chicago."

32. Ames, "Hourly Nursing"; Davis, "Meaning of the Hourly Nursing Experiment in Chicago"; Florence Patterson to C.-E. A. Winslow, 11 November 1926 and 18 November 1926, Winslow Collection, 34: 55; Dorothy Deming, "One Way Out: An Answer to Some Problems of Private-Duty Nursing," *Survey* 56 (June 1926): 377–79.

According to Mary Gardner, unlike the poor, hourly patients had not yet learned to accept "instructive service" without protest; see Gardner's *Modern Problems,* 429, and "Hourly Nursing Service," *PHN* 20 (March 1928): 126–28. Her *PHN* article discusses the

Philadelphia hourly service but fails to note that, as late as 1927, most visits were made to board members. Chronically ill patients were the most frequent users of the services. See Harriet Leek, "The Hartford Hourly Nursing Service," *PHN* 20 (May 1928): 238–40; Louise Tattershall, "Hourly Nursing in Public Health Nursing Associations," *PHN* 19 (August 1927): 397–402.

Patterson and her staff thought the demand for hourly nursing should come from the community—from people interested in securing superior nursing service at a lower rate than private-duty nursing rates. They were apparently content to wait for the community to come to this realization and demand the service. Florence Patterson to C.-E. A. Winslow, 11 November 1926 and 18 November 1926, Winslow Collection, 34: 55; CHA, Minutes of the Nurse Committee, 27 May 1930, and Minutes of the Board of Managers, 28 May 1928, IDNA Collection.

33. Michael Davis, "Nursing Service Measured by Social Standards," *AJN* 39 (January 1939): 38–39.

34. For exact figures, see financial statements in the annual reports of most VNAs. For a rather scattered review of VNA sources of income for 1923 and 1924, see "Policies and Problems of PHN Services," *PHN* 17 (September/October 1925): 483–87, 534–36. The 1927 average earned income was 35 percent; see "Earned Income of 29 Public Health Nursing Associations," *PHN* 21 (April 1929): 216. For voluntary hospitals, 18 percent of income came from tax monies. See Rosemary Stevens, "A Poor Sort of Memory: Voluntary Hospitals and Government before the Depression," *Milbank Memorial Fund Quarterly/Health and Society* 60 (1982): 551–84; NOPHN, *Survey of Public Health Nursing: Administration and Practice* (New York: Commonwealth Fund, 1934), 121–26.

35. "Their Health Is Your Health," Henry Street Nursing Settlement, Annual Report 1934, VNSNY Collection.

36. In 1920, IDNA case percentages were the following: acute/communicable, 38 percent; maternity/newborn, 49 percent; chronically ill, 11 percent; and not requiring nursing care, 2 percent. Thirty percent of all babies born in Boston were cared for by IDNA nurses. The maternity service included "periodical visits from the nurse before confinement, and bedside care for mother and baby after confinement." In much of the city, nurses were available for attendance with the doctor at the time of delivery. See IDNA/CHA, Annual Reports 1912, 1927, 1928, and 1929, IDNA Collection.

In 1912, nurses could not find the patient at home 3 percent of the time, but this small number (2,240) was of little financial impact. By the 1920s, the number reached a much more alarming level—twenty thousand (7%) of three hundred thousand visits per year.

37. Haven Emerson, *Philadelphia Hospital and Health Survey* (Philadelphia: Philadelphia Hospital and Health Survey, 1930), 27; Elizabeth Fee and Theodore Brown, "Public Health at the Crossroads," *AJPH* 89 (November 1999): 1645–48. For statistics on CHA, see Beard, *Nurse in Public Health*, 61–75. In 1920, the case percentages were acutely ill, 59 percent; chronically ill, 11 percent; and maternity/newborns, 30 percent.

38. Emerson, *Philadelphia Hospital*, 453. Thirty-eight percent of the patients dismissed by the visiting nurses in 1928 were acute medical cases, 24 percent maternity, 21 percent newborns, 11 percent chronically ill, and 5 percent other.

39. Lillian Wald, *House on Henry Street* (New York: Henry Holt, 1915), 26–27; Lillian

Wald, "The Treatment of Families in Which There Is Sickness," *AJN* 4 (December 1904): 427–31; Wald to Schiff and Loeb, 8 May 1894, Wald Collection, NYPL; Annual Report 1927, 12–13, VNSP Collection. In 1922, IDNA nurses made 44,500 visits to 3,133 chronically ill patients, or fourteen visits per case. Nine percent of patients were chronically ill, but these patients received 17 percent of visits. See Annual Report 1923, 20, IDNA Collection. Estimates of expenditure for chronic illness in Philadelphia in the late 1920s were based on 3,000 chronically ill patients, 54,000 visits to the chronically ill, a total patient case load of 33,814, total visits of 257,850, a charge of $1 per visit, and a total budget of $244,326. See Minutes of the Board, 19 November 1926, VNSP Collection. Wald wrote about the problems created by the chronically ill caseload as early as 1893; see Wald to Schiff, 3 November 1893 and 8 May 1894, Wald Collection, NYPL.

40. Notes from a discussion of the problems chronic illness posed for staff of Henry Street Settlement are reviewed in "The Care of Chronic Patients," *PHN* 20 (January 1928): 26–27. See also Charles Rosenberg, *The Care of Strangers: The Rise of America's Hospital System* (New York: Basic Books, 1987), 237–61, 287–309.

41. The heading of this section is from "When Is a Chronic a Chronic?" *PHN* 20 (June 1928): 300–301.

42. "When Is a Chronic a Chronic?"; Marguerite Wales, "Round Table on Care of the Chronic," *PHN* 20 (September 1928): 464–66; Helen Stevens, "Round Table on Chronically Ill," *PHN* 20 (October 1928): 541–42; Ernest Boas, "The Proper Institutional Care for Patients with Chronic Disease," *Modern Hospital* 19 (November 1922): 405–8.

43. Sophie C. Nelson, "Round Table on Chronically Ill," *PHN* 20 (October 1928): 541; Sophie Nelson, "Study of Chronic Illness: Carried on by the John Hancock Mutual Life Insurance Company and the Community Health Association of Boston, Mass.," *PHN* 21 (November 1929): 577–78; Karen Buhler-Wilkerson and Susan Reverby, "Can a Time-Honored Model Solve the Dilemma of Public Health Nursing?" *AJPH* 74 (October 1984): 1081–82.

44. Naomi Deutsch, "Round Table on Chronically Ill," *PHN* (1928): 541–42.

45. Doris Roberts and Jan Heinrich, "Public Health Nursing Comes of Age," *AJPH* 75 (October 1985): 1162–71; Stevens, *In Sickness and in Wealth,* 140–70.

46. Roberts and Heinrich, "Public Health Nursing"; Pearl McIver, "Some Findings of the National Organization for Public Health Nursing's Survey of Public Health Nurses of Significance to State Health Administrators," *Public Health Reports* 49 (September 1934): 1081–90; NOPHN, *Survey of Public Health Nursing.*

47. The MLI-funded study is published in NOPHN, *Public Health Nursing Care of the Sick: A Survey of Needs and Resources for Nursing Care of the Sick in Their Homes in 16 Communities* (New York: NOPHN, 1943). Over the years, the ability of VNAs to meet the needs of the chronically ill varied. This was especially true during so-called nursing shortages—especially during World War II. See Hortense Hilbert, "Community Nursing Service during Wartime," *AJPH* 33 (March 1943): 239–45; Agnes Fuller, "More about Wartime Adjustments," *PHN* 37 (May 1945): 242–45; Victoria Grando, "Making Do with Fewer Nurses in the United States, 1945–1965," *Image: Journal of Nursing Scholarship* 30 (second quarter 1998): 147–49.

48. U.S. Department of Health and Human Services, Division of Nursing, *Survey of*

Community Health Nursing, 1979 (Washington, D.C., U.S. Department of Health and Human Services, 1982), 14, 22; Louise Tattershall, *Public Health Nursing in the United States: January 15, 1931* (New York: NOPHN, 1931); U.S. Division of Public Health Nursing, "Nurses in Public Health" (Washington, D.C.: DHEW, 1962); "How Many Public Health Nurses?" *PHN* (January 1941): 21–22. Sixty percent of these nurses were employed by VNAs.

49. "'Planning for Chronic Illness,' A Joint Statement of Recommendations by the American Hospital Association, American Medical Association, American Public Health Association, and American Public Welfare Association," *AJPH* 37 (October 1947): 1260. For an interesting analysis of this topic, see Ernest Boas, "The Chronically Sick," *AJN* 37 (February 1937): 137–43.

50. The next major meeting (after the Commission on Chronic Illness forum) to consider the complex problems of caring for the sick at home was in December 1952. Eighty-two experts (board members, physicians, nurses, health officers, and representatives from a variety of social agencies) from the United States, Canada, and England came together for a five-day retreat. Ruth Hubbard, Director of the VNSP, described the announcement of the meeting as a long-awaited call—to Mecca. The meeting (the Arden House Conference) was funded by the MLI, United Defense Services, and the National Foundation for Infantile Paralysis. Predictably, health departments persisted in their reluctance to care for the sick, physicians sought analysis that depended on using their heads more than their hearts, and nurses simply maintained hope for the future. National League for Nursing, *Report of Conference on Public Health Nursing Care of the Sick at Home* (New York: National League for Nursing, 1953).

51. Ernest Boas, *The Unseen Plague: Chronic Illness* (New York: J. J. Augustin, 1940), 75–76. Boas was the Medical Director of Montefiore Hospital for Chronic Disease in New York City in the 1920s.

52. Ibid, 76.

53. Herman Somers and Anne Somers, *Doctors, Patients, and Health Insurance: The Organization and Financing of Medical Care* (Washington, D.C.: Brookings Institution, 1961), 57–58, 172–81; Joan Lynaugh and Barbara Brush, *American Nursing: From Hospital to Health Systems* (Cambridge: Blackwell, 1996); Stevens, *In Sickness and in Wealth*.

Chapter 9: Home Care Becomes the Fashion—Again

1. The saying "still knocking" was used by the Chicago VNA in its Annual Report 1954, VNAC Collection. Home care persisted; see, for example, Larry Branch, "Home Care Is the Answer: What Is the Question?" *Home Health Care Services Quarterly* 6 (1985): 3–11. See also Annual Report 1955, VNAC Collection; Minutes of the Board of Managers, Boston VNA, "Report of Meeting in NYC," December 1942, IDNA Collection. See annual reports for Chicago and Philadelphia VNAs for the 1920s and 1950s.

Chronic illness now had an official definition that included permanent impairment, nonreversible pathology, and the need for long-term care/rehabilitation; see *National Conference on Chronic Disease: Preventive Aspects of Chronic Disease Proceedings* (Baltimore: Commission on Chronic Disease, 1952), 14.

In 1928 the nurses had 37,048 cases and made 269,542 visits. In 1954, they had 12,981 cases and made 148,847 visits. All information from the VNAC Collection. Boston's VNA had a similar experience; see, for example, Annual Report 1952, IDNA Collection. From the late 1920s to the mid-1950s, its annual caseload decreased from forty-three to twenty-one thousand. By the 1950s, much of the visiting nurse's work consisted of giving injections. For example, in 1954 Chicago's nurses gave 79,012 injections; the next year they gave 82,588. Until agencies switched to "dry sterilized" syringes, boiling syringes and needles in the patient's home was very time consuming.

2. The Boston VNA's annual reports document deficits in the 1930s ranging from $861 to $36,000. In the 1940s, wartime demands reduced the staff by twenty-five nurses. The VNA had a surplus of $8,000 in 1942 but in most years experienced deficits of $6,000 to $53,000. By the late 1950s, deficits were reaching $70,000 to $100,000, with total expenses of about $600,000. See, for example, Minutes of the Board of Managers, May 1942 and April 1959, IDNA Collection. The Annual Report of the Associate Director, Minutes of the Board of Managers, 1959 (IDNA Collection), lists various proposals. In the 1950s, Chicago's expenses were similar but its deficits smaller, ranging from $1,000 to $13,000; see Annual Reports for the 1950s, VNAC Collection. Losses reached $42,432 in Philadelphia and $73,000 in Boston by the late 1950s; see Annual Reports, VNSP Collection and IDNA Collection.

3. For an excellent overview of the financial status of VNAs, see Alice deBenneville, "Financing Voluntary Public Health Nursing Services," *Nursing Outlook* 2 (April 1954): 185–86. Chicago's VNA received about 42 percent of its income from interest and dividends, 12 percent from the community chest, and 23 percent from patients. See Margaret Wiles, "Financing Nursing Care of the Sick at Home," *Nursing Outlook* 9 (November 1961): 687–91; J. F. Follmann, Jr., *Health Insurance, Nursing, and Home Care: A Study* (Chicago: Health Insurance Association of America, May 1959), E, 1–14.

The percentage of income from departments of welfare varied across the country from less than 5 percent to over 50 percent. See Annual Report 1955, VNAC Collection. In the 1950s in Boston, the cost per visit was $3.50 and reimbursement per visit to welfare patients was $2.50. Welfare cases formed 7 percent of the caseload. In 1952, 32,987 (20%) of 155,221 visits were made to welfare patients. In 1953, Boston reported billing for $70,975 but collecting $49,615 for welfare patients. Boston had various contracts with hospitals, but most were financed by grants or as special projects, and all were short lived. See Boston VNA, Annual Reports for 1952 to 1958; Minutes of the Board Meetings, June 1952, January 1953, March 1953, December 1953, January 1954, October 1954, and March 1957, and combined board and staff meeting, November 1952; Annual Report of Associate Director, 1959; all in IDNA Collection. For Philadelphia, see Minutes of the Board, 4 February 1955, VNSP Collection. The Department of Public Welfare paid for 17 percent of patients, and 10 percent were paid for by other contracts. See "A Report of a Study of the Effects of the Termination of Metropolitan Nursing Contracts," *PHN* 43 (May 1951): 185–89; "How One City Is Handling the Termination of MLI Contracts," *PHN* 44 (July 1952): 377; Linn Brandenburg, "Financing Voluntary Public Health Nursing Agencies," *PHN* 40 (August 1948): 393–98.

4. See all sources cited in n. 3; Ruth Adams and Iva Torrens, "Home Nursing Care for

Veterans," *Nursing Outlook* 4 (September 1956): 497–99; "Income and Expenditures of Public Health Nursing Agencies in 1954," *Nursing Outlook* 4 (October 1956): 577–81.

5. Margaret Klem, "Is Prepaid Nursing Care Possible?" *AJN* 44 (December 1944): 1154–60. *Prepaid* was a term used to describe both health insurance and prepaid group practice plans such as the Health Insurance Plan of Greater New York. For an excellent discussion of this topic, see Janna Dieckmann, "Great Expectations for Health Insurance: Nursing and Prepayment, 1935–1965" (unpublished paper); Wiles, "Financing Nursing Care," 687–91.

6. Boston VNA, Annual Report 1952; Minutes of the Board, June 1952 and March 1953, and combined board and staff meeting, November 1952—all in IDNA Collection; Minutes of the Board, 4 December 1954 and 4 February 1955, VNSP Collection. Philadelphia established a contract with Blue Cross in 1958, but reported that only four patients had used the service; see Minutes, Board of Managers, 21 August 1958, VNSP Collection. See also Franz Goldmann, "Nursing in Health Insurance Plans," *PHN* 40 (August 1948): 405–8.

7. J. F. Follmann, Jr., *Voluntary Health Insurance and Medical Care: Five Years of Progress, 1952–1957* (Chicago: Health Insurance Association of America, 1958); Alma Haupt and Helen Conners, "Nursing in Medical Plans," *AJN* 51 (June 1951): 357–58; Ruth Houlton, "Prepayment Plans for Nursing," *PHN* 31 (March 1939): 152–53; Mary Sullivan, "Nursing in a Blue Cross–Blue Shield Program," *Nursing Outlook* 3 (December 1955): 644–47; David McAlpin, "Prepayment Plans—Applicable to Nursing?" *PHN* 32 (December 1940): 748–50; Margaret Klem, "Nursing Care in Prepayment Medical Care Organizations," *PHN* 37 (August 1945): 412–19.

Many historians have examined the development of health insurance in the United States. See, for example, Odin Anderson, *The Uneasy Equilibrium: Private and Public Financing of Health Services in the United States, 1875–1965* (New Haven: Yale College and University Press, 1968); Ronald Numbers, "The Third Party: Health Insurance in America," in Morris Vogel and Charles Rosenberg, eds., *The Therapeutic Revolution: Essays in the Social History of American Medicine* (Philadelphia: University of Pennsylvania Press, 1979), 177–200; Dan Fox, "Policy and Epidemiology: Financing Health Services for the Chronically Ill and Disabled, 1930–1990," *Milbank Quarterly* 67 (suppl. 2, 1989): 257–87. For nursing's official views, see Committee of the American Nurses' Association and the National Organization for Public Health Nursing on Nursing in Medical Plans, "Guide for the Inclusion of Nursing Service in Medical Care Plans" (pamphlet), 1950.

8. William Shepard and George Wheately, "Visiting Nurse Service: Community Asset for Every Physician," *JAMA* 149 (June 1952): 554–57. The MLI's two-and-a-half year effort to encourage community support for visiting nursing is described in Ruth Fisher and Alma Haupt, "Community Effort to Provide Visiting Nurse Service," *Nursing Outlook* 1 (January 1953): 46–48.

9. Frances Lewis, "Can We Insure Nursing?," *RN* 14 (November 1950): 32–35, 61–62, 64; Houlton, "Prepayment Plans," 152–53; "Health Plans By-pass Nursing," *RN* 14 (September 1951): 41, 82. For an excellent overview, see Marian Randall, "Nursing in Health Service Plans," *PHN* 36 (June 1944): 311–17.

10. Margaret Klem, "Nursing Opportunities in Medical Care Insurance," *PHN* 43 (January 1951): 8–16; Marion Randall, "Home Nursing in the Health Insurance Plan of

Greater New York," *PHN* 39 (February 1949): 167–70. Apparently forgotten was that the MLI and some VNAs had briefly experimented with prospective payment plans in the past. The VNAs found them most advantageous, while the MLI quickly withdrew its support for this approach. See, for example, Henry Street Settlement, Minutes of the Nurse Committee, 27 July 1917, VNSNY Collection. The MLI paid the VNSP a "lump sum" to care for all policyholders requiring care in the 1940s. When the VNSP decided to discontinue this arrangement in 1946, it expected a return to per-visit payments would result in a loss of revenue; see Minutes of the Board of Managers, June 1942, June 1943, and February 1946, VNSP Collection.

11. These plans included the Group Cooperative of Puget Sound, Kaiser Foundation Health Plan, Inc., and Group Health Insurance, Inc., all in New York. See J. F. Follmann, Jr., *Medical Care and Health Insurance: A Study in Social Progress* (Homewood, Ill.: Richard Irwin, 1963), 303. See also Ruth Taylor, "What We Learned from the EMIC Program," *PHN* 41 (May 1949): 263–69.

12. Associated Hospital Service of New York Hospital–Sponsored Blue Cross, *Report of a Study Concerning the Feasibility of Providing Visiting Nurse Services Following Hospitalization for Blue Cross Subscribers* (New York: Associated Hospital Service of New York, 1957); "Blue Cross Pilot Study of Home Nursing Benefits," *Nursing Research* 5 (June 1956): 44; "Blue Cross Reports on Study of Home Nursing Benefits," *Nursing Research* 7 (February 1958): 22.

13. Sullivan, "Nursing in Blue Cross–Blue Shield," 644–47; Marie Phaneuf, "Use of Visiting Nurse Service Shortens Stay, Releases Beds," *Hospital Topics* 36 (December 1958): 21; Henry Bakst and Edward Marra, "Experience with Home Care for Cardiac Patients," *AJPH* 45 (April 1955): 444–50.

14. Follmann, *Health Insurance*, i–v, 26–27.

15. Franz Goldmann and Marta Fraenkel, *Patients on Home Care: Their Characteristics and Experience* (New York: Council of Jewish Federations and Welfare Funds, 1959), 24–33.

16. Wiles, "Financing Nursing Care," 687–91; deBenneville, "Financing," 185–86; Mabel Reid, "What Does Public Health Nursing Cost?" *PHN* 40 (August 1948): 399–404.

17. Reid, "What Does Public Health Nursing Cost?"; Minutes of the Board, 4 December 1953, VNSP Collection; Boston VNA, Minutes of the Board, April 1958, May 1959, and November 1960, IDNA Collection. The Boston VNA's board of managers meeting reported 7.3 visits per nurse per day. This caused a great deal of alarm. In 1956, nurses were making a much more acceptable 8.2 visits per day. See Board of Managers Minutes, May 1956, IDNA Collection. In 1960, the number of visits per day dropped to 7.2, and the board decided to conduct a time study comparing the Boston VNA's practices with those of other leading VNAs. It found little difference and concluded that staff's use of time was within "reasonable limits" and the overhead costs were justified. See Board of Managers Minutes, January 1961, IDNA Collection.

18. VNSP, "The Role of the Visiting Nurse in the Field of Geriatrics and Long-Term Illness" (paper presented at the Geriatric Symposium, Veterans Administration and Peninsula Academy of Medicine, 24 September 1953); Minutes of Board Meeting, 2 November 1961 and 7 February 1961; all in VNSP Collection.

19. Prudence Priest and Virginia McCann, "Home Care for Mrs. Murphy," *AJN* 57 (December 1957): 1578; Ruth Hubbard, "Philadelphia Home Care Plan" (unpublished paper presented at the Detroit VNA Annual Meeting, 9 March 1955), 2; Follmann, *Health Insurance;* DHEW, *Coordinated Home Care Programs: 1964 Survey* (Washington, D.C.: U.S. Government Printing Office, December 1966), Publication No. 1479.

For reports on the opening of Philadelphia's coordinated home care program, see Report by Ruth Hubbard, 3 March 1950, 201, and Minutes of Board of Managers, 5 October 1951, VNSP Collection. The VNSP decided to expand program branches in 1954. See Annual Reports 1949 and 1951/52; Minutes of Board of Managers, 6 April 1956, 7 February 1956, 1 November 1957, 6 March 1959, and 1 May 1959; all in VNSP Collection. For a review of these developments by the coordinator of the program, see Anna Minkoff, "A Community-Based Home Care Program," *Nursing Outlook* 2 (October 1954): 516–18. For an excellent discussion of coordinated home care programs, see Janna Dieckmann, *Caring for the Chronically Ill: Philadelphia, 1945–1965* (New York: Garland Press, 1999), 110–45. The VNSP was so determined to conduct this experiment and so confident that it would prove its worth that, when unable to raise start-up funds, the board used undesignated principal to finance the new enterprise. See also Eleanor Mole, "Patient Goes Home—So What!" *PHN* 39 (January 1947): 30–34; Harriet Frost and Margery Overholser, "Referral of Patients for Home Nursing," *AJN* 46 (May 1946): 329–32; Alice Kresge deBenneville, "Joint Planning for Hospital and Community Nursing Service," *PHN* 39 (January 1947): 26–29.

20. Josephine Goldmark, *Nursing and Nursing Education in the United States* (New York: Macmillan, 1923). For further discussion on combination organizations, see Karen Buhler-Wilkerson, *False Dawn: The Rise and Decline of Public Health Nursing, 1900–1930* (New York: Garland Press, 1989), 209–215; NOPHN, "Census of Public Health Nurses, 1937," *PHN* 29 (1937): 648; NOPHN, *Public Health Nursing for the Sick: A Survey of Needs and Resources for Nursing Care of the Sick in Their Homes in 16 Communities* (New York: NOPHN, 1943).

21. See the DHEW, U.S. Public Health Service, Division of Nursing's *Nurses in Public Health* (Washington, D.C.: U.S. Government Printing Office, 1962), 6, and *Nurses in Public Health, January 1968* (Washington, DC: U.S. Government Printing Office, 1968), 12, 41. See also Department of Public Health Nursing, National League for Nursing, *Progress Report on Combination Services in Public Health Nursing* (New York: National League for Nursing, 1955).

22. Frances Rauch, President, Community Nursing Service of Philadelphia, Annual Report 1968, VNSP Collection; Leon Hirsch, Martin Klein, and Gertrude Marlowe, *Combining Public Health Nursing Agencies: A Case Study in Philadelphia* (New York: National League for Nursing, 1967).

23. Rauch, Annual Report 1968; Hirsch et al., *Combining Public Health Nursing Agencies;* National League for Nursing, "Income and Expenditures in Public Health Nursing Agencies—1958," *Nursing Outlook* 8 (May 1960): 274–77; Wiles, "Financing Nursing Care," 687–91.

24. For examples of handbooks on home care over the years, see *A Hand-Book of Nursing for General Use* (Philadelphia: J. B. Lippincott, 1886); Lee Smith, *Nursing in the Home*

(Buffalo: World's Dispensary Medical Association, 1919); Lyla Olson, *Improvised Equipment in the Home Care of the Sick* (Philadelphia: W. B. Saunders, 1928); Mary Wright Wheeler, *The Practical Book of Home Nursing* [formerly *The Amateur Nurse*] (New York: New Home Library, 1933); Emma Louise Mohs, *Principles of Home Nursing: A Text-Book for College Students* (Philadelphia: W. B Saunders, 1931); Lona L. Trott, *American Red Cross Textbook on Red Cross Home Nursing* (Philadelphia: Blakiston, 1942); and I. J. Rossman and Doris Schwartz, *The Family Handbook of Home Nursing and Medical Care* (New York: Random House, 1958). See also Follmann, *Health Insurance*, 25–28, 31–34, and Appendix E, "Availability of Services for Nursing Care of the Sick at Home."

25. U.S. Public Health Service, *A Study of Selected Home Care Programs* (Washington, D.C.: U.S. Government Printing Office, 1955), Publication No. 447. For an excellent discussion of the development of long-term home care, see Dieckmann, *Caring for the Chronically Ill.*

26. U.S. Public Health Service, *Study of Selected Home Care;* Dieckmann, *Caring for the Chronically Ill.* One of medicine's earliest experiments in home care began as a teaching exercise in 1930 at Syracuse University College of Medicine. See Frode Jenson, H. G. Weiskotten, and Margaret Thomas, *Medical Care of the Discharged Hospital Patient* (New York: Commonwealth Fund, 1944). The American Medical Association study is in Report by the Committee on Indigent Care, "Organized 'Home Care Programs' in the United States," *JAMA* 164 (May 1957): 298–305. See also Marcus Kogel and Marta Fraenkel, "Home Care Comes of Age," *Hospitals* 27 (April 1953): 49–51, 70–71. Franz Goldmann's comment is in his "Home Care for the Needy and Medically Needy," *JAMA* 148 (March 1952): 1085–87.

27. Ernest Boas, *The Unseen Plague: Chronic Disease* (New York: J. J. Augustin, 1940), 3–18; Dieckmann, *Caring for the Chronically Ill,* 1–34, 273–80; Howard Rusk and Michael Dacso, "The Dynamic Approach to the Care of the Chronic," *Hospitals* 29 (January 1955): 63–65.

28. By 1964, the official definition of a coordinated home care program was a program that was "centrally administered, and through coordinated planning, evaluation, and follow-up procedures, provides for physician-directed medical, nursing, social, and related services to selected patients at home." DHEW, *Coordinated Home Care Programs: 1964,* 1.

29. John D. Thompson, "Nursing Service in a Home Care Program," *AJN* 51 (April 1951): 233–35; Goldmann, "Home Care for the Needy"; E. M. Bluestone, "The Principles and Practice of Home Care," *JAMA* 155 (August 1954): 1379–82.

30. Sidney Shindell, "A Method of Home Care for Prolonged Illness," *Public Health Reports* 65 (May 1950): 651–60. Michael Davis claimed that many of the speakers at a conference on the problems of old age (University of Chicago, 1930) "thought that improvement in the care of the aged is largely a problem of individualization." See I. M. Rubinow, ed., *The Care of the Aged: Proceedings of the Deutsch Foundation Conference, 1930* (Chicago: University of Chicago Press, 1931), 47.

31. Rose DeLuca, "Discussion: Home Care Programs," *AJPH* 43 (May 1953): 603–4.

32. *Charter for the Aging* (Albany: Office of Special Assistant, Problems of the Aging, 1955), 252. This source claimed that the rather confusing term *home care* had become popular during the past decade.

33. The term *back-to-the-home movement* is from Harriet Frost and Theresa Lynch, "From Hospital to Home Nursing," *AJN* 48 (November 1948): 684–86. Montefiore's program is discussed in "Home Care at the Montefiore Hospital (Editorial)," *AJPH* 39 (February 1949): 224–25; Bluestone, "Principles and Practice," 1379–82. See also Martin Cherkasky's "A Community Home Care Program," *AJN* 49 (October 1949): 650–52, and "The Montefiore Hospital Home Care Program," *AJPH* 39 (February 1949): 163–66; Thompson, "Nursing Service," 233–35; Lucille Notter, "Coordinating Home Care for Persons with Long-Term Illness," *PHN* 39 (December 1949): 602–8; George Silver, "Social Medicine at the Montefiore Hospital—A Practical Approach to Community Health Problems," *AJPH* 48 (June 1958): 724–31.

Of course, most VNAs had provided a wide range of services for years. For example, the Detroit VNA added physical therapy services in 1926, occupational therapy in 1933, nutrition and mental health in 1938, homemaking help in 1955, and medical and social work consultation in 1956. See Emilie Sargent, "Visiting Nurse Service," *Public Health Reports* 75 (December 1960): 1140–42.

34. See all sources cited in n. 33; DHEW, *Coordinated Home Care Programs: 1964;* Henry Markley and Jacob Brauntuch, "How Home Care Works for a City of 48,000," *Hospitals* 32 (June 1958): 35–38; Louis Udall, "Philadelphia Plan for Home Care of the Chronically Ill Person," *JAMA* 152 (July 1953): 990–93; Marcus D. Kogel and Alexander W. Kruger, "New York City's Long-Range Program for Extending Hospital Care into the Home," *Hospitals* 24 (January 1950): 35–38, 62–65; David Littauer and I. J. Flance, "Home Care Made a Place for Itself: Report on Three-Year Program Indicates," *Modern Hospital* 89 (August 1957): 78–81, 83–86; David Littauer, "Hospital Care . . . at Home," *Hospitals* 28 (November 1954): 74–77; Judith Abramson, "Bellevue Hospital's Home Care Transfer Program," *PHN* 42 (September 1950): 513–19; Priest and McCann, "Home Care for Mrs. Murphy," 1578–80; Hospital Council of Greater New York, *Organized Home Medical Care in New York City: A Study of Nineteen Programs* (Cambridge: Harvard University Press, 1956).

CHCPs were officially defined as *centrally administered;* using *coordinated planning, evaluation, and follow-up* procedures, they provided for *physician-directed* medical, nursing, social, and related services to *selected patients* at home.

35. U.S. Public Health Service, *Study of Selected Home Care;* Follmann, *Health Insurance,* 55–64 and Appendix G. Specialized programs were set up for patients with tuberculosis, polio, congestive heart failure, and strokes; see DHEW, *Coordinated Home Care Programs: 1964,* 5.

36. Follmann, *Health Insurance.*

37. Ibid., G-5. Other explanations for hospitals' renewed interest in home care included cost of additional hospital construction, increasing costs of medical care, growing medical needs of the chronically ill and aged, and the rising cost of hospital care.

38. The American Medical Association study is described in David Unterman, Arther C. DeGraff, and Henry E. Meleney, "Study of Home Care in a Municipal Hospital," *JAMA* 140 (May 1949): 152–55. Shindell's comment is in his "Method of Home Care," 651–59. Various reports claimed the actual cost of home care as one-third to one-fifth the cost of keeping a patient in the hospital. See also Phaneuf, "Use of Visiting Nurse Service," 21.

39. Phaneuf, "Use of Visiting Nurse Service," 21; DHEW, *Coordinated Home Care Programs: 1964*. The argument about lack of savings by hospitals is made in Goldmann, "Home Care," 1087. Differences in the methods, services, and techniques of home care were said to make cost analysis highly theoretical. See Arnold B. Kurlander, "Problems of Organized Home Care for the Long-Term Patient," *Public Health Reports* 69 (September 1954): 823–28; American Medical Association, "Organized 'Home Care Programs,'" 298–305.

40. See, for example, Kogel and Kruger, "New York City's Long-Range Program," 63; Goldmann, "Home Care."

41. Kogel and Kruger, "New York City's Long-Range Program," 63; Goldmann, "Home Care."

42. Follmann, *Health Insurance*, 34–39, 55–63, and Appendix G, 18. The example used to illustrate the potential problem of reimbursement was that, if charges were based on the cost per patient, the length of stay might increase, especially if charges were prepaid.

43. Follmann, *Medical Care*, 62, 290–317; DHEW, *Coordinated Home Care Programs: 1964*, iii. The claim about the prototype for Medicare was made by John W. Cashman, Chief, Division of Medical Care Administration, U.S. Public Health Service, DHEW.

44. DHEW, *Coordinated Home Care Programs: 1964*, 1, 11–29. Twenty-seven states had no programs, four states had five or more programs, and the Northeast (New York, Pennsylvania, New Jersey, and Delaware) had twenty-one programs. Hospitals (27 voluntary and 11 public) initiated and supported the largest number of programs, followed by health departments (14), visiting nurse associations (9), other community organizations (6), and medical schools (3). The caseloads of forty-two CHCPs were less than fifty patients; only eight cared for more than a hundred patients. See also Claire Ryder and Bernard Frank, "Coordinated Home Care Programs in Community Health Agencies— A Decade of Progress," *AJPH* 57 (February 1967): 261–65; Abbie Watson, "The Public Health Nurse in the Richmond Home Care Program," *PHN* 44 (May 1952): 263–67.

45. DHEW, *Coordinated Health Care Programs: 1964*, 32–72.

46. Ibid., iii, 1, 72; Phaneuf, "Use of Visiting Nurse Service," 21; Abramson, "Bellevue Hospital's Home Care"; Shindell, "Method of Home Care"; Thompson, "Nursing Service"; Martin Cherkasky, "A Community Home Care Program," *AJN* 49 (October 1949): 650–52; Udall, "Philadelphia Plan for Home Care"; Littauer and Flance, "Home Care Made a Place for Itself"; Unterman et al., "Study of Home Care in a Municipal Hospital"; Kogel and Kruger, "New York City's Long Range Program."

47. DHEW, *Coordinated Home Care Programs: 1964*, 1–13; American Medical Association, "Organized 'Home Care Programs,'" 300.

48. American Medical Association, "Organized 'Home Care Programs,'" 300.

49. DHEW, *Availability of Services for Nursing Care of the Sick at Home* (Washington, D.C.: U.S. Government Printing Office, 1964), iii, 1–6, 19.

50. Some health agencies required additional legislative authority to collect fees, but increasingly this became a reality. DHEW, *Availability of Services*, 9–10, 13–14; "Bedside Nursing Care by Official Agencies," *PHN* 33 (June 1945): 333–34; Charles Haymen and Leona Perkins, "Home Care of the Sick in Relation to Other Public Health Nursing Programs," *Nursing Outlook* 2 (September 1954): 472–74; Committee on Nursing Adminis-

tration of the NOPHN, "Administration of Home Nursing Care of the Sick by Health Departments," *PHN* 37 (June 1945): 339–42; E. M. Holmes and Paul W. Bowden, "Health Department Activities in the Field of Chronic Disease," *AJPH* 41 (July 1951): 812–18; National League for Nursing, "Income and Expenditures," 274–77; Wiles, "Financing Nursing Care," 687–91.

51. The heading for this section is from Goldmann and Fraenkel, *Patients on Home Care,* 29. Mabel Reid's report is in her "Nursing in New York City's Home Care Program," *Nursing Outlook* 2 (October 1954): 530–32, 591–93, and 2 (December 1954): 647–49. See also Jane Holliday, *Public Health Nursing for the Sick at Home: A Descriptive Study* (New York: VNSNY, 1967).

52. Holliday, *Public Health Nursing.*

53. Goldmann and Fraenkel, *Patients on Home Care.*

54. A. E. Benjamin, "An Historical Perspective on Home Care Policy," *Milbank Quarterly* 71 (1993):129–66.

55. For an excellent study of the ANA's support of health insurance, see Dieckmann, "Great Expectations." See also Follmann, *Medical Care,* 447–67; Eugene Feingold, *Medicare: Policy and Politics: A Case Study and Policy Analysis* (San Francisco: Chandler Publishing, 1966), 96–156; Wilbur Cohen, "The Forand Bill: Hospital Insurance for the Aged," *AJN* 58 (May 1958): 698–702; Margaret Klem, "Medical Care Insurance and the Nurse," *AJN* 46 (June 1946): 387–90.

56. Howard Berliner, "The Origins of Health Insurance for the Aged," *International Journal of Health Services* 3 (1973): 465–73.

57. Richard Harris, *A Sacred Trust* (Baltimore: Penguin Books, 1966), 71–103; Follmann, *Medical Care,* 457–59. See also C. P. Trussell, "Senators Pursue Concerns of Aged," *New York Times,* 15 November 1959; Damon Stetson, "McNamara Issues a Credo for Aging," *New York Times,* 10 December 1959; C. P. Trussell, "Democrats Urge More Aid to Aged," *New York Times,* 9 February 1960.

58. Transcripts from these countrywide hearings documented general support for home care and outlined some of the difficulties it posed. See, for example, *Hearings before the Subcommittee on Problems of the Aged and Aging of the Committee on Labor and Public Welfare, United States Senate* (Washington, D.C.: U.S. Government Printing Office, 1959), 58, 61, 71, 455, 480, 503, 602, 612, 674, 771, 781, 885, 893, 898, 930, 937, 939, 940, 941, 945, 1141, 1143, and 1165. For survey responses related to home care, see Subcommittee on Problems of the Aged and Aging, *A Survey of Major Problems and Solutions in the Field of the Aged and Aging* (Washington, D.C.: U.S. Government Printing Office, 1959), 552–60. The subcommittee also issued another report, Senate Report No. 128, *Action for the Aged and Aging* (Washington, D.C.: U.S. Government Printing Office, 1961).

59. *Hearings* (1959), 3, 602, 674, 1165.

60. Subcommittee on Problems of the Aged and Aging, *The Aged and Aging in the United States: A National Problem* (Washington, D.C.: U.S. Government Printing Office, 1960), Report No. 1121, 3.

61. Ibid., 96–97.

62. See, for example, T. R. Marmor, *The Politics of Medicare* (New York: Aldine Publishing, 1973); Robert Ball, "Perspectives on Medicare," *Health Affairs* 14 (winter 1995): 62–

72; Paul Starr, "Transformation in Defeat: Wilbur Cohen, 'Reflections on the Enactment of Medicare and Medicaid,'" *Health Care Financing Review* (suppl., 1985): 3–11; "The Changing Objectives of National Health Insurance, 1815–1980," *AJPH* 72 (January 1982): 78–88; Harris, *Sacred Trust.*

63. Special Committee on Aging, United States Senate, *The 1961 White House Conference on Aging: Basic Statements and Recommendations* (Washington, D.C.: U.S. Government Printing Office, 1961), 37, 39–40, 60; White House Conference on Aging, *Aging in the States: A Report of Progress, Concerns, Goals* (Washington, D.C.: U.S. Government Printing Office, 1961), 66, 98–99, 123, 130–31.

64. Home care appeared in various House and Senate versions of the King-Anderson, Mills, and H.R. 6675 bills. See, for example, Feingold, *Medicare*, 150–51.

65. Benjamin, "Historical Perspective," 134–39. For an overview by the ANA's spokesperson, see Julia Thompson, *The ANA in Washington* (Kansas City: ANA, 1972). See also the following government publications (Washington, D.C.: U.S. Government Printing Office): *Hearings before the Subcommittee on Federal and State Activities of the Special Committee on Aging, United States Senate* (1961), 1 December 1961, 1606, and 23 October 1961, 275, 319–320; *Health Services for the Aged under the Social Security Insurance System: Hearings before the Committee on Ways and Means, House of Representatives* (1961), 1: 3, 4, 34, 36, 57, 66, 136, 186, 254, 271, 275, 303–6, 319–20, 329, 333, 425, 435, 445, 668, 696, 803, 1078, 1114, 1150, 1180, 1204, 1243, 1321, 1376, 1401, 1446, 1451, 1452–54, 1484, 1491, 1532–33, 1573, 1779, 1961, 1986, 2005–6, 2014, 2079, 2104, 2201, 2257–58; *Medical Care for the Aged under the Social Security Insurance System: Hearings before the Committee on Ways and Means, House of Representatives, on HR 3920* (1964), 141–42, 169, 190, 203, 386–88, 478, 491–92, 511–12, 649–50, 655, 731, 748, 796, 1075–76, 1083–84, 1124, 1138, 1333–34, 1384, 1560, 1561, 1669, 1817, 1820, 1823, 2218, 2336, 2411, 2414, 2425, 2429, 2436, and 2437; *Hearings before the Subcommittee on Federal, State, and Community Services of the Special Committee on Aging, United States Senate* (1964), 22–26, 38, 70, 98, 99, 206, 240, and 321; *Medical Care for the Aged: Executive Hearings before the Committee on Ways and Means, House of Representatives, HR 1* (1965), 5, 134, 153, 154, 156, 158, 161, 168, 183–84, 311–12, 341–45, 373–75, 377–78, 380, 399, 436, 447–52, 457, and 470; *Social Security: Hearings before the Committee on Finance, United States Senate, HR 6675* (1965), 4, 5, 11, 12, 24, 94, 182, 184, 201, 290–92, 309–11, 358, 387, 397, 423, 424, 454, 457, 461, 462, 486, 555, 563, 567, 568, 585, 608, 696–97, 699, 1064, 1102, 1146, 1193, 1217, 1220, and 1252.

66. For a review of nursing's efforts to determine the cost of home care, see Mabel Reid's "What Does Public Health Nursing Cost?" and "The Visiting Nurse Service of New York Is Cost Conscious," *PHN* 44 (March 1952): 137–39. The nurse who developed the cost-accounting method was Goldie Levenson. Elsie Griffith, retired executive director of the VNSNY, claimed in February 1990 (interview with the author) that the National League for Nursing, especially Inez Hayes, was instrumental in assuring home care's inclusion in the Medicare legislation. Brahna Trager, who for years was the editor of *Home Health Care Services Quarterly* and an influential participant in many home care developments in Washington, confirmed the league's involvement behind the scenes in home care's inclusion in Medicare; interview with Brahna Trager, September 1991.

67. Feingold, *Medicare*, 306; Benjamin, "Historical Perspective"; Brahna Trager (in-

terview, September 1991) claimed that she introduced the terms *in-home services* and *home health care,* probably in a report to the Senate.

68. Report of the President, Miss Elizabeth Madeira, 4 November 1965, VNSP Collection. The heading of this section is taken from this report.

69. For an analysis of Philadelphia's history, see Sam Bass Warner, *The Private City: Philadelphia in Three Periods of Its Growth* (Philadelphia: University of Pennsylvania Press, 1991). In 1998, Miss Madeira celebrated ninety years of life and fifty years as a board member of the VNSP. Both the person and the organization are alive and well, ready for the certainty of home care's future challenges.

70. Warner, *Private City;* Claire Ryder, Pauline Stitt, and William Elkins, "Home Health Services—Past, Present, Future," *AJPH* 59 (September 1969): 1720–29. Ryder was a physician whom some (not the nurses) referred to as the "mother of home health care." See also Eva Reese and Mabel Reid, "Medicare: A Visiting Nurse Service Interpretation," *Nursing Outlook* (June 1966): 33–34; Ruth Spurrier, Ruth Johnson, and Ann Manussen, "Challenges to Nursing in Medicare," *AJN* 65 (November 1965): 68–75.

71. Karen Buhler-Wilkerson, "Home Care the American Way: An Historical Analysis," *Home Health Care Service Quarterly* 12 (1991): 5–17.

72. Ryder, "Home Health Services"; Benjamin, "Historical Perspective." In 1967, 1,753 agencies were certified to participate in the Medicare program. Of these, 549 were VNAs, 93 combination agencies, 939 health departments, 133 hospitals, and 39 other. See National Association for Home Care, *Basic Statistics about Home Care* (Washington, D.C.: National Association for Home Care, 1997), 2, 5.

73. This dialogue began with a special issue of *Inquiry* devoted to home care. "Special Issue: Home Care," *Inquiry* 4 (October 1967).

74. Brahna Trager, "Home Health Services and Health Insurance," *Medical Care* 9 (January/February 1971): 89–98.

75. Ibid.

76. Ibid.; Brahna Trager, "Home Health Care and National Policy," *Home Health Care Services Quarterly* 1 (spring 1980).

Epilogue: The Future of Home Care

1. William Weissert, "Home Care Dollars and Sense: A Prescription for Policy," in Daniel Fox and Carol Raphael, eds., *Home-Based Care for a New Century* (Malden, Mass.: Blackwell Publishers, 1997), 121–33.

2. J. D. Arras, ed., *Bringing the Hospital Home: Ethical and Social Implications of High-Tech Home Care* (Baltimore: Johns Hopkins University Press, 1995); J. B. Wood and Carol Estes, "The Impact of DRG's on Community-based Service Providers: Implications for the Elderly," *AJPH* 80 (1990): 840–43.

3. Fox and Raphael, *Home-Based Care,* 1–4; C. M. Tauber, *Sixty-five Plus in America* (Washington, D.C.: U.S. Government Printing Office, 1992); A. E. Benjamin, "An Historical Perspective on Home Care Policy," *Milbank Quarterly* 71 (1993): 129–66; Virginia Conley and Mary Walker, "National Health Policy Influence on Medicare Home Health," *Home Health Care Services Quarterly* 17 (1998): 1–15; Richard Hegner, "Medicare Cover-

age for Home Care: Reining in a Benefit out of Control" (National Health Policy Forum Issue Brief No. 694, 1996), 2–7.

4. Hegner, "Medicare Coverage for Home Care," 3.

5. Ibid., 1–10; Fox and Raphael, *Home-Based Care;* Edward M. Campion, "New Hope for Home Care?" *New England Journal of Medicine* 333 (November 1995): 1213–14; W. G. Weissert, "Home and Community-Based Care: The Cost-Effectiveness Trap," *Generations* 10 (1985): 47–50; Christine Bishop and Kathleen Carley Skwara, "Recent Growth of Medicare Home Health," *Health Affairs* 13 (fall 1993): 106.

6. H. Gilbert Welch, David E. Wenneberg, and W. Pete Welch, "The Use of Medicare Home Health Care Services," *New England Journal of Medicine* 335 (August 1996): 324–29.

7. Ibid.; Benjamin, "Historical Perspective"; Fox and Raphael, *Home-Based Care;* Karen Buhler-Wilkerson, "Home Care the American Way: An Historical Analysis," *Home Health Care Services Quarterly* 12 (1991): 5–18.

8. Nora Super Jones, "Access to Home Health Services under Medicare's Interim Payment System" (National Health Policy Forum Issue Brief No. 744, 1999).

9. Ibid.; William A. Sarraille, "The Home Health Industry Faces Mounting Allegations of Fraud," *Journal of Long-Term Health Care* 17 (summer 1998): 10–18; Mary Mundinger, *Home Care Controversy: Too Little, Too Late, Too Costly* (Rockville, Md.: Aspen Publications, 1983).

10. See, for example, the National Association for Home Care Web site: http:// www .nahc.org. See also David Hess, "U.S. Cost-Savings Moves in Home Health Care Seem Off Target," *Philadelphia Inquirer,* 21 February 1998, A3; Alice Ann Love, "Medicare Warns Agencies on Home Health Cutbacks," *Philadelphia Inquirer,* 4 February 1998, C1–2.

11. Jones, "Access to Home Health Services." The National Association for Home Care has issued numerous news releases on this topic; see, for example, "Medicare Home Health Care Interim Payment System Is Inherently Unfair to Patients and Providers," 15 July 1998. The theme for the September issue of the association's magazine, *Caring,* was "Medicare Today, Tomorrow—Forever?" See Judith Feder and Marilyn Moon, "Can Medicare Survive Its Saviors?" *Caring* (September 1999): 30–33; Dayle Berke, "The Balanced Budget Act of 1997—What It Means for Home Care Providers and Beneficiaries," *Journal of Long-Term Health Care* 17 (summer 1998): 2–9.

Typical of home care's resiliency, the VNA of Greater Philadelphia kept its mission alive in an environment made extremely difficult by the Balanced Budget Act of 1997. During 1998 and 1999, its nurses made five hundred thousand visits to 15,588 patients and their families. Its comprehensive array of services included hospice, infusion, pharmacy, adult day care, meals on wheels, community outreach, and health-promotion programs. In addition, the association opened VNA House Calls, a collaborative primary care practice in which a team of nurse practitioners and physicians provide care in patients' homes. "Making Strides and Meeting Challenges Marked Year at the VNA," 1999 Annual Report, VNSP Collection.

12. National Association for Home Care, *Basic Statistics about Home Care, 1999* (Washington, D.C.: National Association for Home Care, 1999); Peter Arno, Carol Levine, and Margaret M. Memmott, "The Economic Value of Formal Caregiving," *Health Affairs* 18 (March/April 1999): 182–88; Carol Levine, "The Loneliness of the Long-Term Care

Giver," *New England Journal of Medicine* 340 (May 1999): 1587–90. The National Association for Home Care is once again using the term *private-duty;* see, for example, *Caring Magazine* 18 (October 1999): 51.

13. Charles Rosenberg, *The Care of Strangers: The Rise of America's Hospital System* (New York: Basic Books, 1987); Rosemary Stevens, *In Sickness and in Wealth: American Hospitals in the Twentieth Century* (New York: Basic Books, 1989); David Rosner, *A Once Charitable Enterprise: Hospitals and Health Care in Brooklyn and New York, 1885–1915* (Princeton: Princeton University Press, 1982).

14. Robert Pear, "Annual Spending on Medicare Dips for the First Time," *New York Times,* 14 November 1999, 1, 26.

15. Robert Pear, "Medicare Spending for Care at Home Plunges by 45%," *New York Times,* 21 April 2000, A1, A20. The editorial appeared four days later: "The Plunge in Home Care," *New York Times,* 25 April 2000, A22.

16. Michael Vitez, "A Challenge to Where the Disabled May Receive Long-Term Care," *Philadelphia Inquirer,* 14 June 1999, A1; Robert Pear, "U.S. Seeks More Care for Disabled outside Institutions," *New York Times,* 13 February 2000, A24.

17. C. Eng, J. Pedulla, G. P. Eleazer, R. McCann, and N. Fox, "Program of All-inclusive Care for the Elderly (PACE): An Innovative Model of Geriatric Care and Financing," *Journal of the American Geriatrics Society* 45 (1997): 223–32; Mary Naylor and Karen Buhler-Wilkerson, "Creating Community-Based Care for the New Millennium," *Nursing Outlook* 47 (May/June 1999): 120–27. These issues are constantly being brought to the attention of the public. The *Philadelphia Inquirer* ran a four-part series during the spring of 1998 examining the high cost of living longer. The conclusion was that America must now face the cost of care for millions. Sara Rimer has written several front-page articles in the *New York Times;* see, for example, "Blacks Carry Burden of Care for the Elderly," *New York Times,* 15 March 1998, A1.

18. Anne Somers, "Long-Term Care at Home," *New England Journal of Medicine,* 341 (September 1999): 1005.

Index

Adams, Claudia Green, 81
Addams, Jane, 244n9
Aetna Life Insurance Company, 161, 261n63
The Aged and Aging in the United States:
A National Problem, 198
aides, home health, 53, 67, 187, 199, 208
Aikens, Charlotte, 65
almshouses, 4, 5, 7, 8, 9, 11, 216n12
AMA. *See* American Medical Association
American Hospital Association, 180
American Journal of Nursing, 95
American Journal of Public Health, 190
American Medical Association (AMA), 180,
  189, 191, 194, 199
American Nurses Association (ANA), 199,
  267n29
American Public Health Association, 97, 180,
  189
American Public Welfare Association, 180
Americans with Disabilities Act (ADA), 211
Anderson, Cecile Batey, 95
apothecaries, 3, 46, 227n4
Arden House Conference (1952), 270n50
Armstrong Association of Philadelphia, 83
Armstrong Commission Hearings (1905–6),
  147
Associated Charities Society of Charleston, 69,
  72

Association of Colored Graduate Nurses
  (Philadelphia), 90

Baby Hygiene Association (Boston), 38, 168,
  226n31, 263n11
"back-to-the-home movement," 190, 205,
  276n33
Balanced Budget Act (1997), 210
Baltimore (Md.), 36, 153, 154, 168, 223n5
Banks, Anna DeCosta, 72–77, 79, 81, 88, 95;
  photo of, 73
Beals, Jessie Tarbox, 118
Beard, Mary, 35, 39–42, 151, 167, 176, 263n12; on
  black nurses, 93–94; and CHA financial crisis, 168, 169–70; hiring of, 226nn34,35; photo
  of, 37
Beazley, Ada, 148
Beer, A. E., 34
benevolent societies, 12, 45, 108, 151, 216nn15,19,
  219n43. *See also* Ladies Benevolent Society
Benjamin, A. E., 196
black nurses, 69, 70, 84–97, 234n18, 240n75;
  and black physicians, 241n77; in Chicago, 87,
  91, 93, 241n76; and health departments, 87,
  91, 241nn76,77; 242n85; and Henry Street,
  112, 241n76; and LBS, 76–81; in New York,
  76, 87, 91, 93, 95, 241n76; and nurses' registries, 135–36; in Philadelphia, 85, 87, 89, 91,

Jacques, Mabel, 239n60

Jenks, Helen, 23

Jewish Family Welfare Association, 161

Jewish Philanthropies, 61

John Hancock Mutual Life Insurance Company, 161, 178–79, 261n63

Johns, Ethel, 84, 87, 239nn56,57

Johns Hopkins University, 39, 225n25

Julius Rosenwald Fund, 93, 97, 242n85

Katharine Kent (Gardner), 34, 47–48

Kelley, Florence, 107

Kennedy-Johnson administration, 199

King-Anderson (H.R. 3920) bill, 199

King's Daughters, 27

Kohne, Elizabeth, 11, 218n39

Kraut, Alan, 49, 58

labor unions, 106–7

Ladies Benevolent Society (LBS; Charleston), 1–13, 18, 55, 217n20; and chronic illness, 7–9, 10, 13; funding of, 10, 11–12, 146, 216n19, 218n41; lady managers of, 30, 82; and MLI, 77–81; patients of, 6–7, 8, 9, 70–71, 217n27, 237n38; pensioners of, 8–9, 11; and physicians, 7, 9, 80, 235n26; and race relations, 68, 70–72, 75, 76–81, 236n30

lady managers, 29–42, 56, 101, 146, 224n8; and immigrants, 59, 61; of LBS, 30, 82; and MLI, 150, 151; and non-nurse attendants, 53; and nurses, 29–42, 70–71, 227n46; and physician-nurse relationship, 45, 46, 47; and race relations, 69, 77–78, 81, 86; and social structure, 30–31; and visiting nursing, 29–38, 42, 66, 90

LaMotte, Ellen, 38, 39, 49, 64, 65, 66, 225n25

Landis, Henry, 84, 88, 89

LBS. *See* Ladies Benevolent Society

Lee, Florence (Mrs. Dacre Craven), 21–22, 102, 220nn12,13

Lent, Mary, 66

Levenson, Goldie, 279n66

Lincoln, Leroy, 164

Lincoln School for Nurses (New York), 88, 95

Liverpool (England), 18–21

Loeb, Mrs. Solomon, 100

long-term care, 188–89, 198, 206, 207, 209, 210, 211

Los Angeles (Calif.), 36

Lowell, Josephine Shaw, 104, 245n19

Lowman, Isabel (Mrs. John Lowman), 29, 64, 150

Madeira, Elizabeth, 200, 280n69

Manhattan Health Society, 266n29

Mann, Henry, 237n39

Massachusetts, 47, 185. *See also* Boston

Massachusetts Medical Society and Homeopathic Doctors List, 47

maternity care, 54, 176, 260n45; for blacks, 85; at Henry Street, 111, 112; and hospitals, 163–64; and insurance companies, 77, 155, 156, 159, 161, 259n41; and LBS, 71; prenatal, 38, 156, 158, 257n27, 259n41; in workloads of VNAs, 168, 170, 268n36

McDowall, Helen, 107

McNamara, Pat, 197, 198

Medicaid, 200

Medical Assistance for the Aged, 195

Medical College (Charleston), 3, 70

Medical Society (Charleston), 75

Medicare, 193, 198, 199, 200–202, 205–8, 280n72; costs of, 201, 202, 206–8, 210; home care benefits of, 208, 210–11

mental health care, 262n6, 276n33

Mercy Hospital (Philadelphia), 85, 90

Methodist Benevolent Society (Charleston), 12, 219n43

Metropolitan and National Nursing Association for Providing Trained Nurses for the Sick Poor, 22

Metropolitan Life Insurance Company (MLI), 146–64, 199, 207, 228n13, 229n32, 270n50, 273n10; and blacks, 152–54, 157, 236n33, 258nn29,32; case management by, 155–57, 164; and chronic illness, 150, 153, 155, 156, 157, 158, 159, 163; cost containment by, 155–57, 158; criticism of claims about nursing and mortality rates, 161–62; and IDNA, 225n29, 227n8, 262n2; and LBS, 77–81; and maternity